MYTHS VS. FACTS OF CRYPTOCURRENCY

Myths Vs. Facts Of Cryptocurrency - A Beginner's Guide To Understanding Cryptocurrencies & Blockchain Technology

The birth of Bitcoin in 2009 gave rise to a new worldwide and unrestricted currency and movement that wouldn't fully be recognized until ten years later. With over 10,000 digital currencies and altcoins at the writing of this, it's key to understand what cryptocurrency and blockchain technology are so you know where we're headed for the future and you can be informed accordingly.

Whether you want to invest into crypto or you're seeking more knowledge because it's all you can hear about these days, this book will provide valuable insights. If you do any cursory search you'll find the topic of cryptocurrency is littered with reports of fraudulent scams and illegal activity. Many will even claim it's volatile and extremely risky, and to "avoid at all costs."

However, the goal of this book is not to serve as an investment guide or financial advice, but rather, to educate beginners and those just getting into the investment and technology space of cryptocurrency so they know what to look for, what to avoid, and how to make a sound decision regardless of what investment you choose. Cryptocurrency is not illegal, but there are many who will exploit the ignorant for their personal gain.

Despite these situations, more and more financial, governmental, business, and investor entities are getting involved with crypto. This sheds light on the powerful utility of cryptocurrency beyond its mere monetary value for selling and buying goods. It goes well beyond that and could pave the way for an exciting new future as we shift to Web 3.0.

In ***Myths Vs. Facts Of Cryptocurrency,*** you will discover:

- The Birth of Cryptocurrency
- A Snapshot Into The World Of Cryptocurrency
- Terms & Definitions of Cryptocurrency
- The Value of Cryptocurrency
- Cryptocurrencies Versus Physical (Fiat) Currencies: What You Should Know
- Cryptocurrency Versus the Financial Market
- The Relevance of Cryptocurrencies & Industries
- Top Myths & Hard Facts About Cryptocurrencies
- Mindful Investing In Cryptocurrency
- Blockchain & Cryptocurrencies
- Thoughts On Crypto's Utility: To Use Or Not To Use

I do not want you to fall victim to fraudulent schemes due to naivety or lack of appropriate information or research before delving into the world of cryptocurrency. My hope is that this book will serve as your starting block, a guide into what cryptocurrency is and what it isn't.

The future is a bright one and full of opportunities when you plan and possess the necessary information and trust to make sound decisions and choices.

MYTHS VS. FACTS OF CRYPTOCURRENCY

A BEGINNER'S GUIDE TO UNDERSTANDING CRYPTOCURRENCIES & BLOCKCHAIN TECHNOLOGY

ALEX CAINE

KING OF KINGS

Myths Vs. Facts Of Cryptocurrency: *A Beginner's Guide To Understanding Cryptocurrencies & Blockchain Technology*

KING OF KINGS

King of Kings Publishing
Houston, TX
www.kingofkingspublishing.com
info@kingofkingspublishing.com
@kingofkingspublishing

This book is not meant to serve as financial or investment advice, but merely for educational purposes. Consult with your financial advisor before you invest into any investment, especially one as volatile as cryptocurrency.

ISBN [ebook] - 978-1-956283-02-0
ISBN [paperback] - 978-1-956283-03-7
ISBN [hardcover] - 978-1-956283-04-4
ISBN [audiobook] - 978-1-956283-05-1

This book is not intended to provide financial or investment advice, and serves only as a beginner's guide to understanding cryptocurrency and blockchain technology.

Crypto is highly volatile and you should consult with your financial advisor before investing into any asset or liability class regardless of its utility.

This book is not for advanced crypto investors and will not provide beneficial investment advice or guidance to a seasoned investor.

CONTENTS

THE BIRTH OF BITCOIN

The birth of Bitcoin in 2009 brought about the rising popularity of cryptocurrency in the present time. Cryptocurrency is one of the most discussed topics in almost every space. It has gained a lot of recognition from many individuals and many organizations, with an increasing number of investors buying digital assets. In addition, companies are investing hugely in cryptocurrency and designing their own digital currency that is acceptable for purchasing their goods and services.

The secret is that the value of cryptocurrency is in its scarcity. The more the scarcity in cryptocurrency, the more the increase in its value. There have been controversies on the value of money and physical currency, and this is because cryptocurrency shares almost similar currency characteristics. These controversies are what will be brought into the limelight in this book.

The use of cryptocurrency is not only limited to just buying and selling coins but there are more secrets to be revealed about its uses in different aspects of the financial sector.

Despite the increase in the cryptocurrency trend, there have been lots of reports on scams and illegal activities. Of course, this doesn't make the technology illegal, but hackers use the vulnerability of the cryptocurrency to carry out fraudulent activities.

Many have little to no knowledge about what to do before investing in cryptocurrency. As a result, they have become a victim of this fraudulent act because of their naivety, lack of necessary information, and thorough research before investing in cryptocurrency. This book will give a complete guide into what cryptocurrency is about and how to invest without making a big loss.

Furthermore, this book is embedded with truths that counter the myths about cryptocurrency, guides to investment, the relevance of cryptocurrency to industries, and the type of wallet to use that can safeguard your digital assets.

1

A SNAPSHOT INTO THE WORLD OF CRYPTOCURRENCY

Cryptocurrency is the new wave of technology in this present age. Who could have thought about cryptocurrency as an investment and a means of income in the last decade?

The cryptocurrency rose steadily in the last few years, and it has now emerged as one of the biggest technologies to solve a financial-related problem in this present age. There is also a popular notion about cryptocurrency as being the future, meaning the cryptocurrency is the present and the future. A crypto trader who invested in Bitcoin in 2017 will have a substantial increase in value, as the rate of Bitcoin increased greatly within the space of five years. However, cryptocurrency is a digital, or virtual, a currency used to purchase online goods and services and carry out online transactions.

The word "crypto" refers to the various encryption algorithms and cryptographic techniques that safeguard entries, such as elliptical curve encryption, public-private key pairs, and hashing functions. Cryptocurrencies are systems used to secure payment, denominated in terms of virtual tokens. These tokens are represented by ledger entries internal to the system. Many companies have started issuing their own token, which can be used in exchange for goods and services that the company provides.

Cryptocurrency is a means of exchange, just like money. However, cryptography secures cryptocurrency. This helps keep it protected and makes it nearly impossible to create counterfeit or double-spend. Many cryptocurrencies are modeled based on a decentralized network of blockchain technology. The most popular and most valuable cryptocurrency is Bitcoin. Others include Ethereum, Litecoin, Satoshi, Altcoin, PeerCoin, NameCoin Ripple, Stellar, Dogecoin, Cardano, Tether, Solana, Binance Coin, etcetera.

According to CoinMarketCap.com, a market research website, more than ten thousand different cryptocurrencies are traded publicly. Cryptocurrency continues to skyrocket, raising money through initial coin offerings or ICOs. According to CoinMarketCap.com research, the total value of all cryptocurrencies as of August 18, 2021, was more than $1.9 trillion, down from April's high of $2.2 trillion. The total value of all Bitcoins was summed at about $849 billion.

BLOCKCHAIN

Blockchain is the foundation for most cryptocurrencies in the market. It is entirely different from cryptocurrency, but it is the basis for most cryptocurrencies being founded. It is a system used to record information in a way that makes it difficult and impossible to hack, alter, or fraud the system, through a decentralized system.

A blockchain is a type of distributed ledger technology that is duplicated and distributed across the entire network of computer systems on the blockchain in which transactions are recorded with a cryptographic signature called the hash.

The use of blockchain is not limited to finding solutions to financial problems in terms of cryptocurrency but in other aspects, like securing medical data, cross-border payment, personal identity security, supply chain, anti-money laundering tracking system, logistics monitoring, and lots more.

The technological product of blockchain is cryptocurrency. Conversely, blockchain secures cryptocurrency from theft and prevents third parties from intruding, which makes it a decentralized system, free from the government or any central authorities. That it is decentralized and not issued by any central authority or the government, it can exist out of government control. This means that the government or central authority does not have the power to interfere or manipulate the system and structure of cryptocurrency.

The problem associated with the centralized system of the flow of money led to the launching of the first cryptocurrency, Bitcoin, in 2009, by Satoshi Nakamoto.

HISTORY OF CRYPTOCURRENCY

The first idea about cryptocurrency started in 1998 when people began to reason about a new payment method, other than using the money system. Wei Dai coined the term cryptocurrency and published a description of "B-Money," an electronic cash system. After, Nick Szabo created "bit gold," an electronic currency system that requires users to complete a proof of work function cryptographically, he put together and published solutions. Hal Finney followed the work of Nick Szabo to create a currency system based on reusable proof.

There was a global economic crisis in 2008 that affected millions of citizens, including America's superpower. The effect of this disaster led Satoshi Nakamoto to create the first decentralized cryptocurrency, Bitcoin, in 2009, which uses the SHA-256 hash function. The idea behind the invention of cryptocurrency was to create a decentralized and international currency where the government and financial institutions would have no control over it.

In April 2011, Namecoin was created to form a decentralized domain name system that would make internet censorship difficult. Soon after, in October 2011, Litecoin was designed, as the first successful cryptocurrency to use scrypt as its hash function instead of SHA-256. Another notable currency in history is Peercoin, which was the first to use a proof-of-work/proof-of-stake hybrid system. Finally, the IOTA combined with IoT (internet of things) was the first cryptocurrency not based on blockchain but used Tangle instead.

Many other currencies have been created, though few have been successful, as they have brought little in the technological advancement of cryptocurrency

However, cryptocurrency has been seen as the future to help people gain financial freedom and protect individual wealth. Therefore, there is a need to examine the current status of cryptocurrency.

CURRENT STATUS OF CRYPTOCURRENCY

Though the beginning of cryptocurrency in the last decade was not really good because people didn't trust the new payment system, the view has changed over time, and the use of cryptocurrency is increasing tremendously. Many organizations and companies have adopted this digital currency and created their own payment for their products and services. One of these companies is Kodak, multinational photography, design, and production company that first created its coin called KodakCoin, and the second is Libra.

Some other companies prefer to allow digital currencies as a means of payment for their product and services. Examples are hotel companies like Casual Hotels, Swiss Hotels, Dodler Grand Hotel, some other five-star hotels, airlines like the US-based Surf Air, tourist parks like the National Tourist Office of

Germany, etcetera. More companies are yet to emerge into cryptocurrency technological innovation.

A higher number of individuals are now accumulating their wealth into the virtual currency system.

There has been a lot of controversy about the adoption of cryptocurrency into the government system. Still, the government is looking at the downside of the effect of cryptocurrency on the system. The internal review service has yet to view cryptocurrency as a form of money, and it is still considered a financial asset or property.

However, there have been lots of arguments and debates over the adoption of cryptocurrency, which will lead to explaining the criticism about the use and adoption of cryptocurrency.

CRITICISM OF CRYPTOCURRENCY

The criticism about cryptocurrency started from its non-acceptance at its inception, but people gradually started to embrace it as a new form of money, anyway.

One of the criticisms about cryptocurrency is the fluctuation in the exchange rate, as the design of many cryptocurrencies ensures a high degree of scarcity. Since market prices for cryptocurrencies are based on demand and supply, the rate at which a cryptocurrency can be exchanged for another currency gives a wide fluctuation. For example, Bitcoin has experienced some rapid increase and decrease in value, climbing as high as $17,738 per Bitcoin in December 2017 before dropping to $7,575 in the following months.

Another criticism is that some economists believe that cryptocurrencies will fade in time or be considered to be a short-lived infatuation. One of the reasons is that Bitcoin is not rooted in any material goods; that is, the cost of producing a Bitcoin is directly related to its market price because of the increasingly large amount of energy required to produce it.

Furthermore, it is a known fact that the cryptocurrency environs are not highly secured, including exchanges and wallets. Even when the cryptocurrency blockchain is secure, there can be an infiltration of hacking from a third party. For example, Bitcoin is the most widely used cryptocurrency. Still, since its existence, there have been cases of hacking and theft during several online exchanges, thereby losing millions of dollars worth of stolen coins.

Nevertheless, amid this criticism, many people still see cryptocurrency as the best means of preserving value against inflation and facilitating exchange outside the influence of central authorities, government, and financial institutions.

MARKET SIZE OF CRYPTOCURRENCY

The market size of cryptocurrency is relative, as most people worldwide use cryptocurrency for one reason or the other. However, the global cryptocurrency market was valued at USD 1.49 billion in 2020 and is projected to reach USD 4.94 billion by 2030, growing at a CAGR of 12.8% from 2021 to 2030.

Various factors that keep driving the growth of the cryptocurrency market are:

- Increase in the desire for operational efficiency.
- The increasing demand for transparency in financial payment systems.
- The demand for remittances in developing nations.
- The attractive opportunities that cryptocurrencies give for market expansion.

THE GROWTH AND EXPANSION OF THE CRYPTOCURRENCY MARKET

The cryptocurrency market has been increasingly in high demand as many people are looking into preserving their wealth and value from inflation due to the transparency in the payment system. This has caused a drive and expansion in the growth of the cryptocurrency market. The cryptocurrency market has better data transparency and independence than banks, financial services, insurance, and other business sectors. Therefore, it is likely to grow and expand more in the future.

Individuals who are in the cryptocurrency market find it easy to send and transparently receive payment. If this same method is adopted in the banking sector, customers' information will be more secure without the fear of being hacked.

Nonetheless, using cryptocurrency in organizations allows stakeholders to share payment transparently, without fearing embezzlement and other related financial crimes. This increases the likelihood that cryptocurrency will be accepted in the future.

It has been observed that cryptocurrency has helped increase the economic scale of countries, especially developing countries, serving as a potential to help emerging economies of developing countries, if accepted. For example, Bitcoin has enabled many individuals and businesses to grow and flourish as a source of revenue. The advent of cryptocurrency in developing economies will expand different areas by making finance and financial activities more accessible.

The use of cryptocurrencies in businesses, like tech companies, hotels, airlines, and other industries, is becoming widespread, allowing virtual currencies as their official mode of payment for their goods and services. This will allow more organizations to consider its use in their operations, which can propel the spread and expansion of the cryptocurrency market.

ADVANTAGES OF CRYPTOCURRENCY

Advantages of cryptocurrency are:

It is Cost-Effective. Bank charges for transactions within and outside the country are relatively high compared to cryptocurrency, where little or no fee is charged for every transaction made. For example, when such transactions are made in Bitcoin, the cost is either negligible or non-existent

Cryptocurrency is a Global Currency. Different countries have their currency; hence sending money from one country to another will cost more due to the exchange rate and takes more time to process. Cryptocurrency serves as a solution to the problem of differential currencies. Furthermore, anyone can send cryptocurrencies to anyone and anywhere globally—it has no geographical limitations.

Everyone has Equal Rights. In the cryptocurrency market, everyone has an equal right to trade, unlike financial institutions where one can give preferences to people with high social class. Therefore, cryptocurrency gives the benefit of equal rights to trade in the market.

Cryptocurrency Makes Transactions Easier. There is no need for third parties in the crypto market, and traders can easily transfer funds directly to their partners without the need for a third party, like a guarantor, credit card company, or bank. Instead, the transactions are secured by using public and private keys and different forms of incentive systems like proof-of-work or proof-of-stake.

Security. Most cryptocurrencies are developed with blockchain technology, like Bitcoin, that provides protection for your coins and values. As a result, there is no fear of a third party infiltrating your personal information.

It is Decentralized. There is no room for central authorities, financial institutions, and banks to manipulate and control the system. Therefore, no government policies or financial rules are guiding or control the usage of cryptocurrency.

DISADVANTAGES OF CRYPTOCURRENCY

Disadvantages of cryptocurrency are:

Cybersecurity Issues. Since the inception of the use of cryptocurrency globally, there have been cases of cyber theft and threat to the security of crypto traders. For example, multiple ICOs got breached, costing investors hundreds of millions of dollars, resulting in the loss of $473 million. However, many fear that they might also fall into this ditch with cryptocurrency.

Price Volatility and Lack of Inherent Value. Volatility is the degree, or speed, of change in prices. In cryptocurrency, many people have problems with how cryptocurrency changes in the twinkle of an eye, making it a high risk for investors. In addition, price volatility can affect its value.

Regulations. There is panic among those that trade in cryptocurrency about its regulation. Since no regulations or policies govern the use of cryptocurrency, there is a tendency to bubble up in the future.

Change in Protocol. There is always an improvement to be made for every technology, which can happen later in the future. For cryptocurrency, there can be a change in the technological makeup, which might affect how things are being done, the value, and the procedure in transactions.

HOWEVER, THE RISE IN THE USE OF CRYPTOCURRENCY IS increasing day by day as more individuals, companies, and even countries are considering adopting it into their systems. The price volatility serves as a downside to its use due to inconsistent price values. Nonetheless, it is still seen to preserve value and increase wealth without the fear of authorities and financial policies affecting it. This explains the big move that cryptocurrency will take in the next decades.

2

TERMS & DEFINITIONS OF CRYPTOCURRENCY

As with the traditional financial system, there are a ton of technical terms that newcomers must acquaint themselves with before investing or trading virtual money in the cryptocurrency world. This is important because cryptocurrency itself is a new phenomenon. And, as such, the market is not without certain terms that are associated with it.

Sometimes, it may be challenging to keep track of these technical words. Conversely, this chapter will provide a comprehensive overview of some of these terms so that your understanding of the market will be enhanced.

You must know that some of these words are composite while some are individual words. Notwithstanding, our focus will be on some selected jargon that you must understand as a beginner.

Below are definitions of a few terminologies.

BLOCKCHAIN

Many investors may recognize the need for a blockchain, but they don't understand what it means. Back in 2014, when the idea of crypto came to the orbits of my concern, I used to think the word was associated with Blackhat Russian Operation. Don't bother to ask why. Then I met John Karl, who is now a friend and an expert crypto investor. He explained the concept in simplified terms, and now I can relate to it, copied with a decade of one-on-one experience on the platform.

A blockchain is a virtual or digital ledger, composed of all transactions made in a specific digital currency, this means that blockchain helps to keep records and secure all transactions made on digital currencies like assets, price and ownership. Transactions are made up of "blocks." Blocks can recreate. When a block reaches its maximum capacity, a new block will emerge, and so forth. Some blocks are designed with limited capacity. However, they are some that are designed with a large market cap.

Transactions on cryptocurrencies are not a hiding thing. It's revealing. Almost everyone in the market can have access to the information on a blockchain. In particular, transactions on Bitcoin are completely public.

It's funny the way many associate Bitcoin with the early days of illegal drugs and firearm deals. But the more prominent, the easier it becomes for people to trace transactions on it, the transparency of cryptocurrencies is easily noticed when you are using centralized trading platforms that use the KYC measures.

Still, there are blockchains where transactions are exclusively private. For instance, in the Monero blockchain, no one can link a transaction to any address because of its nature. You can make use of this platform if you don't want your transaction to be revealed. Because of its nature, it has attracted no fewer users who are interested in making complete anonymous transactions.

At this juncture, I must let you know that there is no central location where ledgers are stored on a blockchain. Rather, the ledger is copied repeatedly on various servers and computers globally. This is why it is often considered to be decentralized.

WALLET

The fact that cryptocurrencies are virtual, or digital, doesn't mean they don't have a place where they are stored. Traditionally, money is kept in your account with a financial institution for many reasons. But in crypto, transactions are made on your wallet account.

A wallet is a place where you store your crypto value for transaction purposes. It is a unique code that represents its address on the blockchain. It is a public address, but within it, there are private keys that determine the ownership of an account, as well as the balance on the account.

Wallets can be in different forms; you can have them on paper, hardware, or software. It is also possible for them to be in another form.

As a place where your coins are stored, your wallet must contain keys, addresses, and seeds that allow you to function effectively.

Basically, a wallet is where you store your digital currency —Bitcoin and others. It is a place where you keep your private keys that can be used to access your public currency address and sign in for transactions. It is a combination of the recipient's private key and your public key that makes a cryptocurrency transaction possible.

There are two main types of wallets. We have the software and hardware wallet. If you are using a mobile phone for your trading, you may need a mobile app. That mobile app is an example of a software wallet. That app helps you store crypto on a third-party server. You may also choose to store your currencies in a destock wallet on your computer.

Besides the software app, there is an alternative known as hardware. Personally, I make use of a hardware wallet to store my currencies. Though they are not free, they help to secure your cryptocurrencies offline so they won't be hacked. I do recommend this type of wallet for people who are not well-versed in coding and technical details to help protect their crypto against hackers.

Hardware wallets act like flash drives or other physical devices. Whenever you want to send crypto from your account, you simply connect the flash drive that contains your wallet to a computer system. After that, enter your PIN-code and send your money. The moment you are done with your transaction, disconnect the flash drive and keep it safe.

Unlike the software wallet, the encrypted information is stored on the private keys on the hardware wallet. The moment your hardware receives the information, it will wait for your confirmation. Ask that it would sign the transaction and send back a digital signature to the software to validate and complete the transaction.

A cryptocurrency wallet can either be cold or hot. The hot wallet allows users or investors to receive and send tokens. In other words, a hot wallet gives you the access as the user to send or receive tokens into or from your wallet.

A hot wallet is one of the most known crypto wallets. One of the main reasons why many investors and users rely on and make use of hot wallets is that its secure transactions. They operate using internet-connected facilities, like computers, tablets, and mobile phones. This makes the wallet less secure.

On the other hand, a cold wallet is the safest alternative for storage. The easiest way to describe a cold wallet is that it is not connected to the internet. Therefore, it stands a fair risk of not being hacked like a hot wallet. This type is often referred to as an offline wallet or hardware wallet.

When you want to invest and be assured that you are in safe hands, try a cold wallet. It is the easiest way to store your Bitcoin, Ethereum, or other digital currencies. Also, this type of wallet requires little knowledge to set up. That is why it is essential for you to learn about safe storage and all the concepts regarding both the cold and hot.

COINS

A coin is a virtual cash or digital currency that is independent of a blockchain or platform. The critical feature of a coin is that of a currency. However, the term is used to describe a crypto asset that you are trading with. Coins are not the same as tokens. In function, they perform different tasks and are used on different platforms.

Unlike crypto tokens, coins are not utility in function. What does that mean? It simply means they can't represent votes within a community nor have the storage capacity on a decentralized cloud storage platform. Instead, a coin has its independent blockchain where it functions. It operates independently within a particular financial system.

Basically, a coin is a means of exchange in the crypto market. It is also used as a store of value with a specific digital economic network. Blockchains are designed to distribute a ledger that verifies and tracks transactions. They have native coins that can be transferred between users of a specific network.

A coin is regarded as a single unit of currency. You can trade with them after an agreement based on the value of the current market requirements has been made. Most times, coins can be exchanged for another coin or a token that belongs to an entirely different blockchain. This can be done either through direct private transfer, which is known as peer-to-peer, or through a crypto exchange. Atomic swaps and decentralized exchanges are the most effective alternatives for trading tokens and coins.

DASHBOARD

You have probably heard about the financial dashboard. There are no clear-cut differences in purpose with the financial dashboard you know and that of cryptocurrency.

A crypto dashboard is an electronic or digital platform where you can monitor as many as possible market caps. This dashboard can be on a desktop or on your mobile, but what is important is its functions. Mainly, its work is to help accept all the cryptocurrency accounts or the coin you have added to your wallet for proper monitoring. It helps you to monitor its ever-shifting value and to manage both your assets and other financial activities. It provides a comprehensive overview of all transactions. In addition, it is an information management tool designed for tracking, analyzing, and monitoring crypto transactions.

MINING

This is probably one of the most common words you will come across. Mining is a process by which new currencies are being invented. It is an electronic system through which new digital currencies are entered into circulation. Also, it is the way new transactions are confirmed by the network and a key component to developing and maintaining a blockchain ledger.

Mining is often costly because it makes use of highly sophisticated tools that solve difficult computational and mathematical questions.

This term is relevantly common among miners—people who solve computationally cumbersome questions through hashing. Miners determine the number of currencies that would be in circulation by the reason of how they can solve those questions.

Though the process of mining is a complex and expensive

one, mining still has some magnetic appeals that attract investors, making them want to spend a fortune to mint currencies.

HASHING

This is almost the same with mining. Hashing is also a process by which miners solve some difficult mathematical problems. But for hashing, miners make use of cryptographic hash functions.

This term is used to capture a process of cryptography where a specific form of data is converted to a distinct string of text.

One interesting thing about hashing is that it does not take into account the size or number of data to be hashed. Any number of data, regardless of the size, can pass through this process.

Hash is designed to perform a function—you can put one data into the hashing algorithm and get an entirely different string. However, it is possible to come out with a new string. When this happens, you may not be able to interpret the input data it represents. Most times, a distinct piece of data will produce the same hash or something similar.

SMART CONTRASTS

Just like other businesses, buying and selling digital currencies involves both the buyer and seller. Therefore, the transaction is complete when both have reached an agreement on the value that should be paid for a particular commodity. Until then, the transaction is not complete.

This is the same case with smart contracts in the crypto market. Basically, smart contracts are programs stored on a blockchain but are released when both the buyer and seller have reached a compromise. They are used to automate an agreement so that both parties involved would know the likely outcome. However, the outcome doesn't include time loss or intermediary's involvement. They can be used to trigger the next line of action and automate a workflow when transactions are completed.

Smart contracts are self-executing contract that contains the terms and conditions of any transactions between the buyer and the seller. These specified instructions are directly written into the lines of code. When an agreement is reached, the smart contract will set the next line of action.

Yet, it is important to know the code that instigates the automated actions it contains. The code and the agreements are decentralized blockchain networks. Primarily, they are meant to control the execution of the transaction and to ensure all transactions are trackable and irreversible.

This platform allows transactions among anonymous investors. They do this without recourse to any legal document, approval from a central authority, or any external enforcement agencies. This has further been proven by some legal academics, who have claimed that smart contracts are not a legal agreement; rather, they are a means of fulfilling or undertaking obligations that are derived from other agreements. In addition, some scholars also argue that the declarative or imperative nature of programming languages influences the legal validity of a smart contract.

PERSONAL HUBS

When you are planning to invest in cryptocurrency, there is a need for you to have a personal email. This personal email will be connected to a HubSpot, which will give you access to send a message from your email to another email. You will use your email to reply to the CRM, send sequence emails, as well as install HubSpot Sales. That will give you access to sales tools in your inbox.

When you connect your email, you have access to sending emails to your buyer using your wallet's email server. In spite of this, some HubSpot tools might require a personal email connection. On your personal HubSpot, you have the liberty to connect multiple personal emails to allow other users to send emails, as well.

MARKET CAP

Another term we need to consider is market capitalization. Market cap is the total value of coins that you want to trade with. As an intending investor or trader, it is essential you know the value of your coin.

Market cap is calculated by multiplication. It is about multiplying the price of a currency with the number of coins that are available for trading. It is the sum of coins that are available for mining. The total of a coin will help determine the price of a given at a particular period in the market.

Don't forget you are dealing with virtual currency and, as such, price is one of the ways to measure a crypto value. Investors leverage the market cap to narrate the story and compare value across digital currencies. This is a critical statistic that shows the growth potential of a currency. It also shows when it is safe to save and when it is not.

IN SUMMARY, THERE ARE MANY CRYPTOCURRENCY TERMS OR jargon that you must be acquainted with before you start trading. However, this chapter has provided the most essential ones if you really want to thrive in the digital currency world.

3

THE VALUE OF CRYPTOCURRENCY

There seems to be so much controversy about the rise of cryptocurrency. Usually, this rise is attributed to Bitcoin being the major digital currency representing cryptocurrency. But, unfortunately, while they are not wrong, they are not right, either. So, it stands that Bitcoin was not at all the origin of cryptocurrency. It might have marked a huge success for the crypto world and may be used as a solid standard, but it isn't its origin.

This chapter intends to take you through an in-depth analysis of the value of cryptocurrency. While it will be unassuming that you don't know any of these values, it will also be presumptuous to state that you are sufficiently equipped as to how cryptocurrency has gotten to be so valued. To ensure this balance in knowledge, I will walk you through the history of cryptocurrency adoption into the financial market.

HISTORY OF CRYPTOCURRENCY'S ADOPTION

The major goal of cryptocurrency is to aid digital currency and fix the problems of traditional currency by ensuring that the holder of the currency is responsible for the currency.

With this goal in mind, the first cryptocurrency market adoption was facilitated by the creation of DigiCash in 1995 by cryptographer and computer scientist, David Chaum. At this point, the value of DigiCash as an acceptable form of cryptocurrency was not known, nor was it examined. To be fair, DigiCash was not seen as a revolutionary form of digital trading, though it laid the foundation for Bitcoin. Therefore, the crash of DigiCash was foreseeable, and well, it did crash.

Nonetheless, DigiCash was not an entirely failed project. On the contrary, it spurred more cryptographers to believe in Chaum's work and his belief in the value of cryptocurrency. Consequently, Nick Szabo's "Bit Gold" and Wei Day's "B-money" provided a realistic template of what digital currency should look like. Also, these platforms were modeled after David Chaum's DigiCash, although with significant improvements. Though these platforms did not gain attention and ceased to exist after a while, their constant reinvention set the value and consequent attention for the creation of Bitcoin in 2009.

Bitcoin, as a form of cryptocurrency, did not skyrocket or hit the market upon release. Like its predecessors, it also suffered several instances of lack of attention. During this period, efforts were made to expose people to Bitcoin through the website opened for it—bitcoin.org. However, all these efforts still paled into insignificance until the earlier part of 2009, when the first codes of Bitcoin were written. It would seem like Bitcoin started gaining traction after its first line of codes was written, but this is not entirely true. It took the belief of a number of investors in Satoshi Nakamoto's work to enforce the platform that is now known as Bitcoin today.

It may seem like most of the cryptocurrency conversation is centered around Bitcoin. This is understandable, as it has been argued extensively that Bitcoin remains the oldest and most popular form of cryptocurrency.

When Bitcoin's value rose from 2009 to 2012, it was

greeted with a lot of criticism and controversy. It was discovered that it could be used for illegal activities without a trace of these activities, especially in the Bazaar Silk Road transactions in 2013. Like most people would say, "Whether bad publicity or good publicity, publicity is publicity," the trail of controversies that accompanied Bitcoin brought with it a number of new investors and visitors. Hence, Bitcoin's price began to skyrocket, and around 2017, 1BTC was sold for $20,000.

I believe taking you through this journey of the introduction of cryptocurrency into the market and the development of Bitcoin has not been fruitless. On the contrary, it shows you what underlined the actual value of Bitcoin as the alpha form of cryptocurrency. Furthermore, the boom of Bitcoin encouraged the rise of other forms of cryptocurrency, like Litecoin, Ethereum, Polka Dot, Cardona, Binance Coin, Stellar, ChainLink, Tether, Monero, etcetera.

Cryptocurrency's value came through gradual development. This gradual development was still not occasioned by sporadic growth in the financial market. Instead, the value of cryptocurrency was established by its unique purpose of indicating the essence of digital currency, and this essence, by early 2009, was grasped by people.

THE VALUE OF CRYPTOCURRENCY

To make this section relatable to you, I have deliberately sourced a conventional definition of value which, according to Merriam-Webster Dictionary, is *"The monetary worth of something."* This definition may seem one-sided. Still, it is needed to expatiate this section on the value of cryptocurrency fully.

CRYPTOCURRENCY AS "A GOLD MINE"

From the history I provided about the rise of cryptocurrency, I think you can infer that the price of cryptocurrency started in pennies, then small amounts of dollars. Gradually, it grew from a small number of dollars—about $0.008 in 2010—to tens of dollars in later years. The shocking boom was when the price skyrocketed to about $20,000 in 2017. Now, I would have gladly mentioned the current rate of 1 BTC or 1 Ether, but these things change, and it would be too assuming to suggest in this chapter that the price will remain the same even as this book is being read in some years to come.

My point? Though there are fluctuations in the price of cryptocurrencies, these fluctuations do not harm the primary value of Bitcoin, Ethereum, Binance, Tether, or Litecoin. Here is the explanation.

If 1BTC falls from $65,000 to $40,000, this fluctuation does not precipitate a constant change in the primary value of the coin, as it always rises to meet this primary value. So, you see, the volatility of cryptocurrencies, if anything, does not still affect their original value. If anything, one may dare to say it protects their value, despite the instability. Therefore, will it be deemed a bold statement that cryptocurrency is a gold mine?

Most people don't understand that the primary value of cryptocurrency is not the current price of 1 BTC or 1 Ether.

What are the significant differences between the Bitcoin that sold for $0.0008 in 2010 and the one that sold for $20,000 in 2017?

When Laszlo Hanyech bought a pair of Papa John's pizza pies with 10,000 Bitcoins in 2010, the sum was just $41. Now, it's more than $3 billion. What changed? Changes in the model of the digital currency? Definitely not. The model has remained the same. Therefore, we can deduce that the original value of Bitcoin, Litecoin, or any other coin that might be added to the list in years to come is the essence of their model in the contemporary world. If another form of cryptocurrency were to come up, as better in model and purpose than Bitcoin, then Bitcoin would take a backseat, even though it is the boss of all cryptocurrencies right now.

In establishing that the value of cryptocurrency as a gold mine is in its model as an essential digital currency, I think it is important to note that I have not presented it as valueless. Cryptocurrency is a gold mine, but that doesn't mean it has too much value to be bought. Gold is a sweet money-fetching resource, but it can be bought. The line of realism is that the value of cryptocurrency is still determined by market forces.

CRYPTOCURRENCY AS A NECESSITY

A necessity is something inevitable. Now, cryptocurrency is something that sounds highly inevitable, and this questions the basis of its necessity in the first place.

For the sake of more precise understanding, let us examine another explanation of value by Dictionary.com as *"The quality of anything which renders it desirable or useful."* Pay attention to the key terms "desirable" and "useful."

The desirability of cryptocurrency is underlined by anonymity. Although this has been heavily criticized because it is a sure way of protecting illegal transactions among terrorists and lawbreakers, it stands that people like that their person is protected, regardless of whether they have committed crimes or not. But this is not the only thing. Cryptocurrency uses a secure ledger with strong cryptography to protect online transactions. This encourages people to purchase the bulk of goods from companies that issue tokens through cryptocurrency platforms.

To the next key term, "useful." We have to be sincere; cryptocurrency has benefited a lot of companies and individuals by transitioning them from mediocre financial situations to a stable financial status. Also, cryptocurrency platforms are beginning to supplement banking organizations, as these organizations are not as decentralized or free as crypto platforms, which allow for free trading from anywhere in the world, and the transactions are as minimal as possible, something that does not collect in banking institutions.

Furthermore, there seems to be a huge benefit for entrepreneurs and small business owners. I honestly didn't understand this until I had a firsthand experience with a friend of mine in Egypt who happens to do financial transactions with some companies in Europe and America. He always complained about how these transactions have been so impeding, so much so that he started considering the possibility of limiting his business coverage. However, with the presence of cryptocurrency platforms came wide financial coverage. In addition, he uses TenX's digital wallet, which helps him route some of his funds to other business investments.

The last analysis on usefulness is the number of jobs created by cryptocurrency. The interesting thing about cryptocurrency is how it has created tons of dynamic ways of rewarding employment opportunities rather than the conventional ones that might not be so rewarding and cannot go around for the entire populace. There are just so many job listings. You have companies calling for adept skills in machine learning, Python, Java, C/C++, Node.js. In fact, from 2017 to 2018, these listings increased to about 194%. Over the years, the creation of more jobs was necessitated by the technical know-how needed to run transactions on cryptocurrency. These jobs include:

Data Scientists. These people are needed to analyze transaction data and uncover what seems like a myth about blockchain technology to everyone who has an interest in cryptocurrency.

Machine Learning Engineers. They will help prevent hackers from attacking cryptocurrency platforms. This will make the platforms secure and user-friendly for users.

Financial Analysts. Millions of people make money from analyzing the best ways to make cryptocurrency investments and how to develop investment strategies against the volatility of these platforms. They offer these services to other people who trade and invest on crypto platforms.

Blogging and Journalism. Journalists and Bloggers make tons of money from just providing information to the public about cryptocurrency, like finding a way to embed in pop culture. Most of the tabloids are laced with countless news of its growth and progress. Therefore, with the growth of crypto platforms have come several websites and blogs dedicated to the single purpose of keeping the community updated about cryptocurrency.

Blockchain Web Developers. There will always be the need for blockchain web developers who can write and maintain codes to allow people to conduct easy transactions on cryptocurrency platforms.

Technical Writers. The history of Bitcoin is incomplete without referencing the White Paper written by Satoshi Nakamoto in 2008. Even though Bitcoin did not boom upon its creation, this paper encouraged a lot of investors to believe in Satoshi's work. As new forms of cryptocurrency arise, there will be a need for technical writers who can adequately convey the purpose of these new platforms to attract investors.

INVESTMENT

We all know the extent of the volatility of cryptocurrency. However, these fluctuations do not still rule out its importance in being one of the best forms of investment in the 21st century. Many people have argued that it is a kind of investment that can yield better profit returns over time. This is why the bulk of those who trade in cryptocurrency with the sole purpose of investing are not doing it for the sake of petty profit returns or a quick one million dollars. They do it for futuristic purposes, and no matter the fluctuations, strategic measures make them smile widely at the end of the day!

WHAT ARE RESTRICTIONS?

On a platform that allows you to trade or conduct transactions twenty-four hours a day, seven days a week, you know you have all the time to yourself to make the best of your investment. Though there are downtimes and technical updates, these are infrequent, at best. But this is not what makes cryptocurrency the best.

The platforms allow you to take advantage of the rising and falling of the market by allowing you to go long or short. While this might not sound like an advantage for many people because of the risk associated with it, the neutrality of this advantage has been proven by the fact that people who decide to go short intend to bank on falling markets, and people who go long do so because they believe the price of particular crypto will fall. This process, therefore, subtracts any form of certainty apart from the one derived from the discretion of traders.

Having walked you through a thorough explanation of the value of cryptocurrency, I believe it's crucial to summarize some of the major points mentioned earlier so they can serve as takeaways for you from this chapter:

- The adoption of cryptocurrency into the financial market was not magical. It was precipitated by laid down plans of people like David Chaum's "DigiCash," Nick Szabo's "Bit Gold," and Wei Day's "B-Money." These cryptocurrency platforms established a foundation for the introduction of Digital Currency

- It will be realistic to state that these former cryptocurrency platforms failed because the world was not ready for it, and people did not see the need or importance of digital currency. Even Bitcoin did not rise immediately upon its creation.
- Therefore, though determined by market forces, such as demand and supply, competition, and regulations, the value of cryptocurrency is still heavily reliant on its model in presenting the essence of digital currency.
- Cryptocurrency takes the world on a speedy global economic revitalization through the creation of rewarding jobs necessary to complement its steady rise.
- The volatility of cryptocurrency is not in question. However, this does not render its value fluctuating, as this fluctuation is, in fact, a huge advantage for people who decide to go short on cryptocurrency platforms. Therefore, managing these fluctuations is best with having risk-assessment strategies in place.
- Finally, the cryptocurrency used to be what was mentioned on fringe websites and what people didn't see as "bankable" upon its start. Now, it's at the center of every financial news source and, in fact, *every* news source. This influence is good enough to say that the value of cryptocurrency will keep it alive for decades to come.

In the next chapter, we will compare cryptocurrency with financial markets and other investments, like stocks and bonds available in these markets. We will also discuss the changes in behavior between cryptocurrencies and the level of their volatility.

4

CRYPTOCURRENCIES VERSUS PHYSICAL (FIAT) CURRENCIES:

WHAT YOU SHOULD KNOW

In a lot of literature on the subject, you will find a discussion of the direct differences between cryptocurrency and physical currencies. In such write-ups, you will find differences like manner of storage, legality, whether the currency is limited or unlimited. Most of the literature will also discuss the advantages and disadvantages of both.

In this chapter, the differences that will be discussed will include those that have been mentioned, but they will be along lines of differences in how each currency is regulated in relation to other currencies of its type. It is worth noting that, when we talk about physical money in this context, we are not just talking about the liquid cash used for transactions. We are also talking about all the forms of digital money that exist in the ledgers of all the institutions that perform those regulatory purposes.

The main issue that people have had with physical money is how its value is tied to its regulatory agencies (i.e. the government and the banks). The value of physical money is so volatile that a wrong policy from the bank or someone in government could cause the owner of the money to be worth a lot less. Thus, to keep their money stabilized, many are now turning to cryptocurrency technology. Many people currently think of it as a better means of carrying out transactions. However, do not just take their words, and mine, for it. Let me explain the differences that exist between them and decide for yourself.

Real money, known as fiat currency, and cryptocurrency both have their operations. Nevertheless, over the past few years, there has been increasing interaction between the two due to the increasing acceptability of crypto.

DECENTRALIZATION

All the decisions on fiat money are carried out based on the directive of the central bank or the equivalent controlling financial institution of the country. All decisions made on physical currencies, either in cash or in the ledger of banks, are centralized, and whatever transactions are carried to be carried are governed and regulated by the government and other licensed financial institutions. On the other hand, it goes without saying that the reason why the cryptocurrency is called that is that, by nature, it is hidden from central control.

As stated previously, cryptocurrency is built on a technology called blockchain, which enables it to use an incorruptible digital ledger to maintain transaction records. Because of this incorruptibility, the ledger itself acts as a trusted source and eliminates the need for a trusted third party. Thus, you can use your cryptocurrency without having a central governing body that makes decisions, enforces them, monitors, and maintains how it should be used.

The decentralized nature also means that there will be no need for third parties, like the government, involved in the decision-making process on how much of the physical money to issue out and how much to keep in its ledgers. This is perhaps the reason why a lot of governments still shun cryptocurrencies and even though those who accept them establish a lot of policies to regulate them.

For most governments, despite the belief that blockchain technology is incorruptible, it can be used for various nefarious purposes, like funding insurgents (as we have seen in Africa in recent times), and otherwise undermining the government.

TRANSPARENCY

Not only are cryptocurrency transactions free from third parties, but they are also transparent. The ability to carry out transparent transactions increases the confidence of its user. All transactions carried out on a cryptocurrency are visible to all users on the cryptocurrency. No information is hidden.

This is different from how the physical money in the banking ledgers is run. The banking ledgers that physical money goes through are closed, isolated, and private. Banks do not give the full user access to all transactions carried out and will only give the public a glimpse of some parts of the records that they manage.

TRANSACTIONS FEES ON CRYPTOCURRENCY VERSUS TRANSACTION FEES IN PHYSICAL MONEY

Blockchain technology and banks offer different services in regulating cryptocurrencies and physical currencies. Therefore, their transaction fees on their products differ. Of course, banks may offer different kinds of services on how best to hold physical money. For example, the money that is held in as credit is moved around by card payments, and so the fee you will be required to pay depends on the card and will not be paid by the user directly. Card fees would be paid to the payment processors by stores. All transactions carried out using this card usually attract fees, and they would be paid after each transaction. The fee attracted most times affects the price of goods and services by making them rise.

There are other fees charged by banks on services like checks and wire transfers. The amount of customers' payments depends on the bank they are using.

The similarity between physical money and cryptocurrency, in this case, is that transactions in cryptocurrency are also charged. Albeit, not in the same way as fees are charged on physical money transactions. For example, using the most popular crypto at the moment, Bitcoin. There are variable transaction fees that Bitcoin attracts. These fees will be determined by Bitcoin miners and users. Transactional fees in the Bitcoin network can also reach as high as $50, but the amount each user pays will be determined by the user. Each user can decide if they are willing to pay a certain amount to complete a transaction.

Bitcoin users are offered a choice, unlike the conventional financial system. The choice also creates an open marketplace where your transactions would not be processed if you set a fee too low.

SPEED OF TRANSACTION

The rate at which transactions are completed on the cryptocurrency and banking platforms differs greatly. The speed of transactions depends on the traffic of the network at a particular time. For example, transactions on the Bitcoin network can be completed within fifteen minutes, or it can take over an hour if the network is congested.

However, there are bank transactions that could take up to three days before they are completed. The time needed for bank transactions to be completed depends on the kind of transaction that you are performing. Most card payments should be completed within forty-eight hours, while wire transfers should not take more than twenty-four hours. If you want to do international transfers using banks ensure that it does not fall on weekends or holidays because transactions would not be processed on such days.

KNOW YOUR CUSTOMER RULES

Because the central bank through the lower banks and other financial institutions regulates fiat money, they often build "know your customer" procedures, which they use to determine the identity of their users. If you want to use any service offered by banks, like bank accounts and banking products, you must provide banks with your means of identification. All banks are legally required to have a record of their customer's identification, not just before they are able to open an account to keep their physical money in and to be able to move it around.

Nevertheless, cryptocurrency services can be offered to anyone with or without them needing to identify themselves. Even in countries where cryptocurrency services are regulated heavily, the user of the cryptocurrency still has some measure of leeway to decide if it will be necessary to identify themselves or not. For example, you can use a cryptosystem without providing any identification. In fact, robots with artificial intelligence can move cryptocurrency from one wallet to another.

EASE OF TRANSFERS

All you need to make use of cryptocurrency-based services is a device that can connect to the internet and a good internet connection. You can carry with you a mobile phone and internet connection. You must have a bank account, mobile phone, and identification issued by the government before you make use of any bank service.

PRIVACY

The information that will be published on a cryptocurrency depends on how the cryptocurrency technology is used. Just to be clear, the privacy of all Bitcoin users rests on the shoulders of the users. The only information that will be displayed on the public record is the amount transacted and the public address of the user. No private information of the user will be displayed, but you can track how a number of Bitcoins move. You cannot determine the identity of ownership of Bitcoin that was bought anonymously.

It is possible to trace the ownership of a Bitcoin to know the identity of the buyer if it was bought on a platform that requires the identification of its users. Any Bitcoin purchased on that exchange will be tied to the owner of the account.

You cannot make use of physical money without the bank being able to trace the source of the transaction, its destination, and what the transaction is for. This vital information is stored on private servers that are kept safe by banks. Any information about physical money must be stored on the private servers of banks and held by the client. The privacy of all the physical money in an economy bank user depends on the security of the banks' (both the central bank and the lower banks) servers.

SECURITY

One of the most important attributes that a financial system must provide in order to regulate physical money is security. However, the security strength of financial systems differs because of the measures involved and the design of the security system itself. A loss of confidence in the government or the banking system can see the value of fiat money drop in a matter of minutes. Thus, the security of a bank account depends on the security measures each user employs and the bank's private servers.

Even though it's not an easy feat to hack private bank servers, it is still possible. You can use measures like strong passwords and two-step authentications to bolster the security of your fiat money.

The security of the cryptocurrency depends on how it is used. The security of a decentralized cryptocurrency can only be breached if a hacker gains access to half of the devices maintaining the network. In Bitcoin, the strength of the cryptocurrency increases with the network, and the amount of security a Bitcoin holder has with their own Bitcoin depends on them. You should store your crypto coins in cold storage instead of hot ones.

APPROVED TRANSACTIONS

There are different types of transactions carried out daily by the banks and other agencies who regulate physical currencies, and the availability of a system that can support a variety of those transactions is important when deciding which financial system to use. The kind of transaction that can be carried out in a bank will be determined by the bank itself. Banks have the right to terminate a physical currency transaction for various reasons. They can also decide to freeze your medium of holding physical money.

Banks monitor the kind of transaction that each customer carries out, and they can decide to deny transactions of unusual items or at unusual locations.

Cryptocurrency-based assets do not nitpick transactions, and users are free to carry out any form of transactions, even if they are illegal (again, necessitating government oversight), especially on platforms that do not require identification. If you are using cryptocurrency-based platforms that require identification, you should ensure that all your transactions are in line with the laws of that country and region.

EASE OF UNDERSTANDING

You do not have to be tech-savvy before you understand the workings of cryptocurrency technology. It is also incredibly easy for cryptocurrency to adapt to new changes. New features and functions can be added, despite the high level of innovation the existing cryptocurrency displays. The adaptability of banks is not so fluid as any changes that will be made to the system must adhere to its regulations. Still, its regulations allow it to maintain its security. It also holds it back from improvements due to its restrictions.

ERROR RESISTANT

The transparency, decentralization, and security provided by the cryptocurrency make it a more trustworthy system than the centralized ledgers that hold your physical money. Any information that has been entered into a decentralized cryptocurrency cannot be altered or tampered with, or even lost due to poor management. The irreversibility of the data on the cryptocurrency makes it a more secure and trustworthy method of value storage than physical currency.

Given that the value of your money is determined by the central bank or the regulatory agency, the bank can decide to open a policy that would devalue the country's currency, causing your money to suddenly be worth a lot less.

CAN PHYSICAL MONEY AND CRYPTOCURRENCY COEXIST?

Although many consider cryptocurrency a significant upgrade to the existing banking system, it is still difficult for many to determine which one is the ideal system to power the future of money. This indecisiveness that many have shown could be because the banking system has been tested and trusted, despite its obvious issues. It could also be because cryptocurrency is a relatively new technology.

Although cryptocurrency is the backbone of many coins (Bitcoin, Ethereum, Dogecoin, etcetera) that are quickly becoming the world's currency, there has been an increase in the use of cryptocurrency in 2021. A reported 880% rise in its usage, according to a 2021 Chainalysis Global Crypto Adoption. One of the reasons for this is because crypto has shown to be an innovative new way that can play a valuable part in the financial future of the world. However, the cryptocurrency system has not yet proven to many, especially governments of nations big enough to make significant changes in cryptocurrency use, to be a reliable banking system. What this means is that, for now, both physical money and cryptocurrency have their advantages, strengths, and weaknesses.

Rather than focusing on their weaknesses in a bid to determine the better system of carrying out financial transactions, a third option might be the solution. The third option here is an interdependence between cryptocurrency and the banking system (i.e. the regulators of fiat money).

Blockchain experts are already working to create a cryptocurrency that combines the function of a public and private cryptocurrency, which can be used by banks and other financial institutions. An example can be observed in the Ripple cryptocurrency that increases the speed of transactions and allows them to perform cross-border banking.

Combining both kinds of cryptocurrency—private and public into a hybrid cryptocurrency—allows cryptocurrency to overcome challenges, like slow transactions, high expenses, and the risk of illegal transactions. If these challenges are solved, it makes it easier to migrate into the existing banking system. The introduction of a cryptocurrency with reduced challenges might remove the perceived competition and allow cryptocurrency and banks to work together. This would impact how the majority views cryptocurrency and the assets it supports.

Cryptocurrency and fiat money do not have to have divergent futures when they can both complement each other. Although, I have compared them to show differences.

5

CRYPTOCURRENCY VERSUS THE FINANCIAL MARKET

If you are new to the cryptocurrencies world but have significant experience with stocks, understanding the primary difference between crypto, stocks, and other traditional investment plans can help decide which one to invest in. Then again, cryptocurrencies have become a mainstream phenomenon in the financial market. Right now, the total value of these digital currencies has swelled up to more than $2. As a result, investors are questioning the place of stocks in the market.

Nearly half of the investors in the market are now considering investing rather than plunking down their money into their stock. The reason for that is not far-fetched. The benefits one can accumulate in crypto are far more than what an investor can ignore. But many of them are swarming into this digital world rush. They are investing with a lot of hope yet little knowledge.

Apart from individual investors, companies are rolling out funds to the digital world, despite its ups and downs. The reason for this is not far-fetched, either. Its rapid appreciation and hype are enough to make one invest in it to expand one's portfolio.

However, a big challenge with crypto is how investors perceive it. This is because most new investors do not know the difference between digital currencies and stocks. As a result, they ended up getting everything twisted.

In this chapter, you will learn the significant difference between the two. This will only help you know certain things about it, but also help guard your decision on how to invest.

To consider the differences, you must understand that the significant difference between the two can be seen in how you value them as an investor. So, as an investor, you probably need to know the value of each of them.

More so, you need to understand that a stock is an ownership interest in a business. This is backed by the company's cash flow and assets of the major. In contrast, digital currencies are not backed by anything at all. Not with cash flow or a company's assets.

Like I have said, stocks have legal backing. They are used by legitimate establishments that are expected to turn to profit. They include physical assets, such as silver, gold, and others as part of their valuation. This makes it easy for investors to calculate the value of a stock in the financial market.

Digital currencies, however, don't have such features. They are not legal tender nor are they backed by any company's assets. Instead, they are valued based on the popularity they get from people.

If you're purchasing digital money, you should understand what you want to buy. More so, you should be able to compare traditional investments, such as stock, which always have a solid long-term record, and others and digital monies.

A QUESTION YOU SHOULD ASK

As an investor, it is important to want what you are actually investing in. Knowing this will help you adequately know well the investment you are about to make. In addition, you will be able to familiarize yourself with it so much that both the risk and the reward are well understood.

Risks and rewards are cogent information every investor must seek to know before they invest their money into the business. This is because they are the critical factors that determine and drive an investment's success.

So, you should ask yourself, "Between virtual currencies and stocks, which should I invest in?" This question will be a driving force that will propel you to seek specific information about the investment. If you don't have adequate information on both the risks and the rewards, you can't make the circulation. You can't be successful in any investment if you have not calculated both. If you have one without these two important things, you are probably investing in something more like gambling.

Below are some basic things you should know about financial markets and virtual currencies.

OTHER FINANCIAL MARKETS

First, it's important you understand that when we talk about another financial market, we refer to bonds, precious metals, and stocks. These are not the same as crypto, even though they have some areas of similarities.

Stocks

Like I have briefly explained earlier, a stock is a *fractional ownership* interest in any business. Don't lose sight of this. The wiggling prices and benefits you can possibly derive from the market might make you overwhelmed. On the other hand, as the owner's stake in the business, stocks enable shareholders to claim a portion of the company's assets and cash flow. The two possibilities substantiate your investment and provide a basis for its valuation and approval.

The Nature of Stocks

Perhaps you have been wondering why stocks rise and fall. Here is the answer: A movement of stock's price is to guarantees the future success of the business. This gives the investors access to the future of the company. While investors may be expecting the best in a short time, the stock price overly depends on the company's ability to increase its profit for the long term. This means that the success of a stock company depends on how stocks rise over a period of time.

Cryptocurrency

Unlike stocks, crypto has no backed hard assets. Though we have stable coins, there is an exception to that cryptocurrency's possibilities. But this may not be the case with popular cryptos such as Ethereum and Bitcoin.

Crypto allows you to perform various functions. You can use it to send money to another investor through smart contracts that automate carrying out the transaction once you have reached an agreement.

The Nature of Cryptocurrency

Like stocks, virtual currencies also have unstable prices. It is not a legal entity because the movement of prices is instigated upon speculation driven by sentiments. This means it's impressive for the price to move when sentiment changes. This may not be too fast, but drastic. Inform your understanding that digital monies are driven by the *idea* that investors will need them more in the future. This is the reason why many investors will buy certain currencies into this wallet then leave it for a while until there are excessive demands in the market. This is done with the hope that someone will buy it for a higher amount in the future. This is what investors refer to as the "greater fool theory of investing."

To make a profit in the crypto world, you need to get someone who will buy more than the price you got it. The market must be more optimistic about it than are.

STOCKS AND CRYPTOCURRENCY. WHAT YOU SHOULD CONSIDER

If your goal is to invest and maximize the potentials in both crypto and stocks, you probably need to take the time to consider your risk tolerance. You should ask yourself if you truly can handle the volatility of both investments. As you know, both are market-based investments with huge volatility rates. So, it's important how well you can respond to losses and gains. More so, you need to ask if you really want to handle the volatility of these kinds of assets.

Below are things you probably have to look into before you plunk your money in any of these investments.

Safety and Risk Management

As market-based investments, both platforms involve some level of risk tolerance. As you have read earlier, the volatile nature of both might not give you access to predict when and how much you should earn within a short period. That means, even as an investor, you might not be able to predict the price of stock nor that of a digital currency due to its unstable nature.

Spot the Difference!

- Unlike digital monies, a stock's level of volatility is higher. This is because there is a high possibility that many stocks can rise up to 100% or more within a financial year and may fall just as quickly.
- Also, as an investor, you can sell a stock and push down the price if you don't like it. Nevertheless, a company has to go out of business for a stock to be worthless.
- The stock market is an effective and robust way to invest with a track record of successful transactions that you can bank on.
- You can own your fund if you don't want to buy individual stocks, like those based on the Standard & Poor's 500, which gained more than 10% per year on average over time.
- Long-term performance on stocks depends on the underlying company's success.

Cryptocurrency

- Unlike stocks, virtual currencies are not backed by a company's assets or cash flow. Rather, it relies on sentiment to push up its price.
- Volatility is not instant; it is drastic, with the price rising or falling fifty percent more in a commonplace year.
- As a result of the lack of legal sanctioning, countries can decide to ban it for life. China and other countries have done that already.
- Because of its relatively new nature, the market is not firmly established yet. So, virtual currencies can't be regarded as an asset class.

- Though stocks are a risky type of business, cryptocurrencies even can be more speculative.

Time Horizon

Another critical thing that must form part of your consideration is the time horizon. This implies the time you need money from an investment. After investing in a business, you probably have the time you are expecting your business to maximize profits for you. This is what is referred to as your time horizon. However, it is essential you know that the shorter your timeline, the safer your asset should be. When it is safer, it will be there when you need it. More so, you should know that the more volatile your asset is, the less suited it becomes for those who may likely need it.

More often than not, experts recommend that investors with risky assets, such as stocks, need at least three years to ride out the volatility.

Below are what you should know about the time horizon of each.

Stocks

They are often volatile but not too volatile, like crypto. It would help if you had it previously. Individual stocks have volatility that is more than portfolio stocks, which are primarily designed to benefit from diversification.

You enjoy more stocks if you are the type who can leave the money alone and refuse to access it, even when you have the urge to do so. This comes with a huge benefit—the longer you leave it, the better.

Stocks can also be more volatile than others. For instance, value stocks or dividend stocks fluctuate less, like growth stocks.

On stocks, you enjoy the privilege of shifting from one difficult (growth stock) to the safer one. Investors approaching their retirement time make use of this because it allows them to tap their money when there is a need to do so.

Virtual currencies

While we consider stocks to be volatile, virtual monies are ridiculously volatile. For instance, in 2021, Bitcoin lost more than half of its value yet regained it back within a few months. Such a volatile nature makes the business unsuited for the short-time investor since they might have to wait a while before maximizing the profits in the market.

With that, we can agree that cryptocurrencies are suited for investors who would like to invest for a long-time purpose. These traders can leave the money tied up and patiently wait until it is recovered. Such traders won't be bothered about the fall of the price since they are in for an extended period. They think of years rather than months and weeks.

Virtual Monies and Bonds

Having learned the similarities between cryptocurrency and stocks, exploring crypto and other financial markets becomes necessary. So, let's take a look at bonds.

Bonds are a loan from an individual to an organization or a government. In simple terms, when a trader purchases bonds from a government or company, the company or government is indebted to that person. What this means is that investors will get an interest on the amount for a period after which the government or company will pay back the entire amount.

However, it is important you know the risk associated with bonds. Prominent is that, if the company goes bankrupt, the investor will lose both the principal amount and interest payment.

Bonds, however, are a bit similar to virtual currencies in terms of risk management and safety. Both cryptocurrency and bonds have good and bad days. Yet, unlike crypto, bonds make it easier for investors to predict the future. They are also regulated, unlike virtual currencies with no central authority that controls their affairs.

Because of the possibility of prediction, risks are minimal. Investors can be sure of a certain amount that should come into their coffer during the time horizon.

Virtual Monies and Forex

Forex means Foreign Exchange. It is a market that attracts traders who invest in foreign currencies. Now, crypto has become a means of exchange globally, and traders who engage in foreign exchange also deal globally. But what seems to be an issue as far as forex is concerned is the unstable economic condition of various countries across the world.

This unstable economic reality has made both cryptocurrency and forex to be grouped together. But then, investors can expect a positive result from forex when the economic condition of the country they invest in is in a good state. Because forex investors receive capital gain on their investment based on the economy of concerned countries, this makes it riskier as compared to virtual currencies.

Virtual Monies and Precious Metal

Undoubtedly, the main reason why people invest in precious metals is to buy jewelry and other such items. So, we can agree that what determines the value of the precious metal is market sentiment.

The risks involved are import taxes, portability, and, last

but not least, security. At the same time, virtual monies and other financial markets don't need physical transfer that will warrant tight securities. This is because, except for precious metals, all other financial markets are digital. This makes it comparatively more accessible for investors who would like to invest.

In summary, cryptocurrency and other financial markets are somehow related in many areas. They all provide financial possibilities for investors who would like to invest for potential profit. They all involve one possible risk or the other. Although people are aware and comfortable with the traditional investment plans, virtual monies are newcomers in the market with many pros and cons that you should be aware of before you venture into the business. Take your time to know each of them and ensure your choice is guarded by the information you have gathered. Choose wisely!

6

THE RELEVANCE OF CRYPTOCURRENCIES & INDUSTRIES

Cryptocurrencies are fast-paced values used as a medium of exchange on digital platforms. It is no news that crypto is a hot cake due to the values attached to it. The goods are numerous; everyone can benefit at one point or the other. It has received exceptional attention from the world. Thus, it is a global deal.

In addition to this, publications and top companies are now becoming devoted to it. Also, legislation from different authorities is now striving to be in alignment with this new financial trend. This makes it easier for them to carry out token issuances; otherwise, they would be left out in the global banking trend.

CRYPTOCURRENCY PACKAGE

Cryptocurrency is a decentralized blockchain-based platform that bestows private ownership to individuals involved. It is not activated by the government or its policies. Therefore, it is free from any manipulations. It is upheld by a peer-to-peer community of computer networks made up of user machines or nodes.

Blockchain is a secured digital database. It is a distributed public ledger that records cryptocurrency transactions and is run by cryptography. It has a network that scrutinizes all information that enters the blockchain, then it confirms and verifies all entries to the ledger, adjusts any change, and brings out errors.

There are multiple currencies in the crypto industry, but I would use the most acknowledged of the currencies—Bitcoin. It is the most popular and valuable coin in the crypto industry. Its popularity could be because of its premier existence in the hands of the great crypto innovator, Satoshi Nakamoto in 2008. Its value has risen tremendously over the years. It is eye-catching and is currently being pursued by a number of internet sellers and retailers.

However, it is prone to price and value fluctuations on rare occasions. For example, before the 2017 Christmas, the price at which these currencies are attained skyrocketed to an unimaginable price, thereby making it difficult to trade crypto. This led to a crash in the market, and a 20% approximation of loss was recorded on its global platform. Therefore, it is expedient; you know that the crypto world deals with speculation. You can lose your money at any time. This is not to scare you but to give you an honest view of what you are getting into.

IMPORTANCE OF CRYPTOCURRENCY

Cryptocurrencies are unique for their role in introducing another form of banking and transaction. This new financial trend is achieved through digital currencies. You are being relieved of using fiat currencies. This does not disregard the importance of fiat currencies, but it offers a convenient way of handling financial interactions. Transactions here are secured, reliable, and swift. It has earned global respect and appraisal. The transactions or exchanges are properly maintained and coded, so you aren't liable to fraudsters. In years to come, cryptocurrencies might replace the traditional banking and transaction system.

CRYPTOCURRENCIES' CONTRIBUTION TO THE GLOBAL ECONOMY

These currencies are virtually incorporated in every country, though some countries like China do not legalize them. Due to its prominence and dedication on the part of the users, cryptocurrencies have made a few contributions to the development of the global economy. It commands trust from its diverse users. Hence, they can interact without being doubtful. Such trust helps in the advancement of international trades and relations, which would result in a symbiotic gain for everyone. Let's examine a few of its impacts.

Economic Acceleration. Crypto contributes both to the individual and society involved. Currently, it is heavily invested in and monitored by distinct institutions. The industry is getting big as the day passes, as a result of the beautiful and mouthwatering benefits. It is a rare opportunity to be financially nourished when properly managed. Bitcoin is a huge factor responsible for the financial increase of some individuals and companies.

In addition to this, it is being held in esteem as many rely on the business as a source of income. It is a nice move if it is embraced and utilized, though it has its odds. The big news is that it is becoming established, and it is gradually being depended upon by the global economy. Once the economy is thriving, debts are reduced.

Transparency. In strengthening any relationship in life, trust is needed, right? This trust is birthed by transparency, and when this happens, all communications will be effective.

Crypto trading is original; it caters to the mental stability of its users. It has blockchain as one of its vital components. The blockchain serves as a security check, and it keeps a detailed record of any transaction. Thus, it is not a tool for manipulation or fraudulent explorations. It is automated and digitized, so it guarantees the truthfulness, unlike manual transactions, which are subject to corruption and mismanagement.

Equal Opportunities. Technology is not restricted to a geographical location; it is designed to be consumed for global progress. Crypto trading is a global meal; therefore, developing countries have their slice of the cake, too. It has increased more job opportunities, merged nations, and continents. It has also reduced the insecurities and discomforts associated with basic transactions and banking.

Compared to the high-interest rates in traditional transactions, crypto offers come in with their high volatility and ease of use. Technology has rapidly assisted in the massive appreciation of the crypto business. It also helps in training its users, which can be of use in other sectors of the country and world.

Minimal Transaction Cost. It is pretty expensive to indulge in financial activities due to the expenses attached to it. One of them is human labor. It is being looked after through wages and salaries. Here, it is a digital business, so those burdens in the traditional setting are not accounted for here. You have no worries about tariffs and rates on your investments and transactions. Thus, operational costs are virtually absent, or low cryptocurrencies are in use as there are no centralized intermediaries to pay. It accounts for positive reviews, as people put more energy into the business. When it is accepted by everyone, global development will be at its peak and trust will be embraced.

Ownership Empowerment. Everybody likes comfort and beautiful benefits from their jobs, businesses, and other things in life. This financial outlook helps its users receive payments in more currencies. Through this act, it develops businesses, either large or small scale.

Can you see how favorable it is for its users? Crypto enriches and empowers them economically and financially. This brings us to the awareness of the massive financial dominance that the crypto business is launching. This financial trend is being networked by blockchain technology, and in little time, people will see the beauty in global investment by exploring the amazing goodies sponsored by cryptocurrencies.

SECTORS AND SEGMENTS INFLUENCED BY CRYPTOCURRENCY

Now that we have seen the relevance of cryptocurrency to the global economy, let's examine its contribution to certain industries and sectors around us:

Finance and Banking Industry. In cryptocurrency, financial activities are a new game. I linked it with the banking industry because of the change it brings to the traditional system. It is a new insight into a limitless and easier way of transacting and interacting. This influence has led many countries to the thoughts of either developing, or thinking of developing, their own cryptocurrency, called the Central Bank Digital Currency (CBDC).

Labor Market. The crypto business is now responsible for the advent of new job opportunities. Man's population seems endless and, deep down, there is a lurking concern about what business or lucrative activities to engage in. Happiness and new directions were brought by this digital business. Its beautiful benefits have attracted both young and old. Therefore, there is enough to go around; you can pursue anyone attached to your strength.

Occupations in this field are marketing manager, financial analyst, account executive, security architect, software engineer, and others. You will be wondering whether these jobs are only for the tech literates, but note that there is provision for people with communication skills (journalism), creativity (content writing), entrepreneurial (marketer/ influencer), and problem-solving skills. This will bring me to the next contribution.

Fountain of Knowledge. You cannot overlook the role technology is playing in the crypto business. Crypto experts are tech-savvy; it is a requirement that anyone with a keen interest in the industry has to have a level of technical literacy. You should be familiar with the workings of technology and digital exploits. The urge to understand the dynamics of this business has provoked a profound interest in global transactions, acute technological utilization, dexterity in solving problems, and many more. Many of the jobs require background experience in technology. In order to get employed, they could decide to have qualitative knowledge in artificial intelligence (AI, C/C++, machine learning, Node.js, etcetera.

Economic Sector. This area is a bit tired of the business sector, and you are now aware of the importance of crypto in an individual's life. Imagine this benefit is synergized; what a wonderful picture it is. If the benefits of cryptocurrency are harnessed, it can have the capacity of holding diverse types of cryptocurrency (global ones), in addition to the national fiat currency. It gives general authority to everyone involved in the business. A man, rich or poor, can perform transactions all around the world in seconds without paying hefty charges to banks.

Government Sector. It can be pretty difficult for the government to adopt it because of the transparency involved. It is a decentralized package that will allow many citizens to have a cryptocurrency wallet on their phones if adopted. It is devoid of government actions like Demonetization, inflation adjustment, etcetera, which regulate the value of fiat currencies. Its decentralized feature makes it free from political manipulations.

However, as inconvenient as it looks for the government, transparency should not be ignored. Government and public records can use blockchain to minimize paperwork and fraud while increasing accountability. It informs them on the financial flow of the country, and it also supports accountability from the government.

Business Sector. As a company, anxiety has no hold on you when you are interacting within the crypto world. This is because of the maximum security it offers—hackers have no tab on whatever business lies between you and your partners. Its decentralized modus operandi helps your effective communication, which builds reliability and mutual interdependence.

In addition to this, it encourages independence in payments and financial transactions, which are now through the internet. You can control your transactions and coffers without deferring to a centralized organization. It also maintains the anonymity of business partners. Here, businesses can explore deft capital raising, which is a more efficient way of raising capital. It is done through a process known as the Initial Token Offerings (ITOs). Customers also have their share in the benefit because they are open to a more convenient way to pay for several services and products—transactions are swift and clear. In the future, businesses will have a lot of innovative and exciting new opportunities in view if they choose to adopt the technology.

WHAT IMPACT DOES CRYPTOCURRENCY HAVE ON BUSINESS?

Cryptocurrency has a strong possibility to disrupt the business ecosystem or the marketplace, just as the creation of the internet. It is having a strong impact on the way customers transact and do business together.

Bitcoin put blockchain on the map. This has produced a significant use case for cryptocurrency. Ever since the first Bitcoin transaction in 2009, several businesses have resorted to using Bitcoin for payment and as a way to pay employees' salaries. This is why blockchain companies are introducing stronger security and productivity in the analytical processes of several industries. This is why I suggest business owners should know about cryptocurrency and blockchain technologies. Businesses should be getting ready to transact with cryptocurrency.

No doubt, cryptocurrency is volatile, irrespective of whether it is widely accepted for investments or transactions. It is determined by how risk antipathetic the user and business is. You must also bear in mind that the worth of Bitcoin could fall tomorrow. There are also operational considerations, like the administrative part, such as tracking trades, payments, and receipts, as well as the custody aspect, which entails how to store the currency securely.

It is true that cryptocurrency lacks regulation. In other words, there is no regulatory body to settle a conflict. Cryptocurrency is deliberately decentralized. As a business or investor, that lack of regulation could pose a potential danger. Any business owner looking to deal with cryptocurrency requires an excellent comprehension of it. Working with a reliable adviser who is knowledgeable to assist a business through it is essential.

A company that accepts Bitcoin as a form of payment could possibly increase its customer base, consequently. It is a universal phenomenon. Hence, the move has the possibility to open up the business to the global marketplace.

Cryptocurrency can also limit payment processing fees. This is because it has lower transaction fees than credit cards. More so, transactions are not temporary, so there are unquestionable chargebacks. If well-managed, transactions are safe, thanks to formidable encryption. It is also a means for an organization to diversify its assets.

Some countries and governments have resorted to blockchain to ease the burden of real estate property sales and title transfers, thus decreasing the bottleneck and making it less easy to falsify records. The automotive industry makes use of blockchain to focus their supply chain management, decrease human mistakes, waste, and extra manpower. Blockchain offers the health care industry data exchange systems that are cryptographically safe and irreversible. This is particularly perfect when working with patient data.

Banks have started to make use of blockchain to decrease fraud by distributing details over the blockchain database, where it is then confirmed on different terminals. They are also making use of blockchain to make money transfer faster and less expensive and to get compliance details on their customers more easily when dealing with regulators. In accounting, blockchain can be used to remove mistakes when dealing with more complicated details from various sources. It also can reduce the quantity of time it requires to finish an audit and decrease fraud.

As regards cryptocurrency, there are some security considerations, benefits, and challenges of Blockchain. Blockchain networks have an auditable operating ecosystem with well-detailed log data that can be examined for compliance, offering security via confirmed transactions, locked contracts on the spread ledger, and an individual set of records that can be seen by all members. It has the possibility to remove reconciliations s=and reproduce ledgers and conflicts over contract terms. Efficiencies are at an advantage since details are always updated, and intermediaries are eliminated from the transaction process.

Cryptocurrency is transforming the business sector positively. With the aid of a reliable adviser, businesses can be ready when the chances come to use them.

7

TOP MYTHS & HARD FACTS ABOUT CRYPTOCURRENCIES

With the increased attention that cryptocurrency has gained in the past few years, there have been lots of misconceptions about cryptocurrency. Many people, including investors and analysts, have asked questions about the advent of cryptocurrency.

There are some myths, rumors, and misconceptions about digital currency, certain coins, and tokens as a result of the significant popularity it has made so far that have manipulated people's perspectives regarding cryptocurrencies. Due to these rumors, some people may have been investing wrongly or have a wrong perception about cryptocurrency, or decided not to invest at all.

Knowing the truth about cryptocurrency is extremely important. Ignorance is dangerous; there is little difference between one who is misinformed and one who is ignorant. The cure to both is the right information. Knowing the myth serves as guided information to knowing the fact about cryptocurrency, which can better aid your decision about crypto assets and monetary investments.

Hence, here are the top 10 myths and truths to counter those myths about cryptocurrency.

MYTH 1: CRYPTOCURRENCY IS USED FOR ILLEGAL OR ILLICIT ACTIVITY

The view that cryptocurrency is been used for illegal activity is one of the myths that has been in existence since the advent of cryptocurrency. The anonymity of cryptocurrency has made some individuals who are criminals hide under the umbrella of cryptocurrency to carry out illegal activities; this criminal activity does not only apply to cryptocurrency alone but also fiat currencies. However, criminals who carry out this illegal activity only hide under the shadow of cryptocurrency. It was the transaction process that was illegal, not the cryptocurrency.

In a broader sense, the intent behind the transactions doesn't make the currency illegal, the same way it is for paper money. Since cryptocurrency does not have a specific security code for a specific person, these criminals get attracted to digital currency.

But the good news is that blockchain contains the user's wallet address that can trace them back using the transaction data and further linked to a real-world identity.

The truth about this is that the environment surrounding the cryptocurrency space is being used for illegal activities. Still, the coins themselves are not illegal because they are built on blockchain technology. However, due to the anonymity and the vulnerability of cryptocurrency, websites and their space are prone to attack from cybercriminals and hackers.

Criminals systematically use several means to get money through cryptocurrency by exchanging Bitcoin for gift vouchers, prepaid debit cards, or iTunes vouchers, and other means. According to analysis, a hacker group called DarkSide shut down the colonial pipeline in 2020 and asked for at least $350 million in crypto ransoms.

Despite these illegal activities around cryptocurrency, the truth remains that cryptocurrency is not a technology that was created for illegal activities but to solve financial problems.

Every transaction in a cryptocurrency is recorded on an immutable blockchain. This means that there are always traces to every trader's identity with the technical know-how.

In a bid to beat this evil act of criminals in the cryptocurrency market, several cryptocurrency companies have brought up law enforcement to track criminal groups, and their responsibility is to analyze the flow of currency.

MYTH 2: CRYPTOCURRENCY DOES NOT HAVE VALUE

Since the wide popularity of cryptocurrency and its acceptance in some countries and organizations, there have been lots of myths about its value, which many believe will fade away soon. Nevertheless, the government has been so concerned about the income tax rule to use for those investing and trading in cryptocurrencies for a long time. There has been confusion on how cryptocurrency will be treated when it comes to payment of tax and regulations of cryptocurrency transactions. This concern and confusion contributed to the myth about the value and the cryptocurrency is a bubble that will burst at any time.

However, the value in cryptocurrency has led international industries to accept cryptocurrency as a form of exchange for goods purchased and services rendered. For example, Bitcoin mining is a process used in creating new Bitcoins, and this process requires lots of energy, and it consumes a lot of electricity because of the mining process, which is the real cost. As new miners join the Bitcoin network, the energy consumption also increases, and the price of Bitcoin rises, as well. Still, the market price and value of Bitcoin revolve around the cost of the mining process. Then again, the more the mining network gets bigger, the more the cost and the more the increase in value.

Many people believe that cryptocurrency is invaluable. The fact to counter this myth is that cryptocurrency *is* valuable. Currency is functional if it is a store of value. This means that it can maintain its value over time. One of the reasons the cryptocurrency is functional is its ability to maintain its value and is at low risk of depreciation.

There are six attributes that determine the value of cryptocurrency and other currencies—scarcity, divisibility, utility, transportability, durability, and the ability to counterfeit. All these are factors that determine its widespread use in an economy. Now, the determining factor of value has changed from the above six attributes.

A prominent Scottish economist, John Law, wrote that currency issued by a government or monarch is not the value for which goods are exchanged but the value for which they are being exchanged. This simply means that the value of currencies is determined by the level of their demand and their ability to pique trade and businesses within and outside the economy.

Fiat currencies and gold are seen as a model for currency, but due to its disadvantages of being immobile, their value is not attributed to their physical attributes anymore. However, there is now an evolution in currency. The digital evolution of money has moved the value of currency from just its physical attribute to its functions in the economy.

The value of cryptocurrency is in the economic system of its demand and supply. The main value of cryptocurrency is a function of its scarcity. The more the supply diminishes, the more the increase in demand for cryptocurrency. The value of the cryptocurrency has continued to increase massively as countries like El Salvador are betting that the use of cryptocurrency will evolve to become a means of exchange in the economy for daily transactions. When this comes to reality, the cryptocurrency will have a much bigger divisibility factor compared to the fiat currency standard unit.

In spite of this, the division can be in the form of eight decimal units called Satoshis. If the price of any cryptocurrency rises over time, traders will be able to buy a tiny fraction of a single coin. They will still be able to take part in transactions involving cryptocurrency, irrespective of whatever price it is. Furthermore, the development of side channels can boost the value of the cryptocurrency's economy. An example of a side-channel is the lightning. The lightning network is a tye of technology that uses a micropayment channels to scale blockchain capability in order to conduct efficient transactions.

Myth 3: Cryptocurrencies are not Secure

Another myth about cryptocurrency is its safety and security. Unlike paper currency, when you can see the bank manager and the financial authorities in charge of your money, it's not the case with cryptocurrency. This has led to some misconceptions about its safety since there are no persons to hold responsible or regulations that control its operation. The truth about cryptocurrency is that it is extremely secure because it was created with blockchain technology.

Many people tend to compare the system operation of paper currency to digital currency. Paper currency is a centralized system where records are kept in one common ledger. This system is easy to attack by criminals. That is, if the central network is being attacked, it can affect every other transaction. This is not the case with cryptocurrency; every record is kept in the form of a block with a blockchain.

Blockchain, as discussed in previous chapters, is the foundation on which most cryptocurrencies are being built, and they are close to impossible to be attacked or hacked by cybercriminals because those blocks cannot be altered or modified after being created. The safety of records is also assured as it is one of the advantages that blockchains have over other technology.

Though, with the level of security in cryptocurrency, there has been a report of various cybercrimes and hackers filtering into the cryptocurrency space. But this incident is a result of a cryptocurrency exchange website's vulnerability and not the cryptocurrency itself. The criminal uses this vulnerability to hack into wallets and other cryptocurrency space. This has caused a lot of investors to worry about the security and safety of their digital assets.

There are ways you can protect your digital assets from fraudsters and hackers. Most importantly is keeping your wallet password safe and not keeping digital currency holding on an exchange.

Since the creation of Bitcoin in 2009, there have been cases of cryptocurrency theft and hackers breaking into this system, but the fact remains that cryptocurrency is a secure technology. Blockchain technology backs up the creation of cryptocurrency. It provides a secure platform and more reason why most organizations use blockchain technology to secure important information, because it is immutable.

Most cryptocurrencies are built with blockchain technology, creating a secure digital ledger for every cryptocurrency transaction. This secured ledger helps to detect hackers and, most times, keep them away.

The backend process of every cryptocurrency transaction is vastly complex, and blockchains help to record these transactions into blocks that are time-stamped. As a result of these complexities, it is especially difficult for cyber attackers and hackers to get through.

Though the environment surrounding cryptocurrency, like the wallet, might not be as secure as cryptocurrency itself, there are several measures to take to make cryptocurrency more secure when making the transaction. The popular security measures known are using the two-stage authentication process while making transactions or entering a username first and a verification code that is sent to your personal smartphone. However, this type of security system might not be enough to keep your cryptocurrency account safe. This is the reason why most companies or individuals probe further to understand cryptocurrency security standards.

Cryptocurrency security standards are a set of security requirements for the systems used for cryptocurrency. This set of security requirements includes cryptocurrency exchanges, mobile, web applications.

In order to increase your cryptocurrency security, it is important to have an information system that has a cryptocurrency security standard. Having this will help to manage and standardize the techniques and perform methods to a particular system for security. This cryptocurrency security standard enables users to make choices and good decisions for investing and purchasing in the right service free of scams and cyberattacks.

This system also helps the customers and investors to make smart decisions when collaborating with the companies. The cryptocurrency security standards have ten points that are being fulfilled while setting up the security system, but these ten security steps are divided into three levels.

The following below are the steps that most companies and blockchain organizations follow in setting up the cryptocurrency security standards:

- Key/seed generation
    - Wallet creation
    - Key storage
    - Key usage
    - Key compromise policy
    - Keyholder grant/revoke policy and procedures
    - Third-party audits

- Data sanitization policy
- Proof of reserve
- Log audits

It is important to implement these security standards while investing in cryptocurrency to protect your asset.

A cryptocurrency security officer helps to safeguard your asset and invest safely without being too overwhelmed with the risk and losses while investing.

Some activities can make a cryptocurrency account a subject of attack by hackers and cybercriminals.

· Leaving cryptocurrency on a single exchange. This makes your account prone to hackers.

· Keeping cryptocurrency locally can be risky as data can be stolen or lost. In addition, local storage is vulnerable to attack as someone can trace down to your transaction and use it to their advantage.

· Your cryptocurrency account can be specifically targeted. Risks like SIM swap can clear the two-way authentication.

· Cryptocurrency can be lost due to natural disasters or accidents. According to research, billions of dollars are the estimated digital currency that has been damaged so far, and many investors usually overlook this cause.

· Not distributing digital assets to beneficiaries can damage it, which means loss of generation wealth. This can happen when a person suddenly dies or has any other complications, which is why it is important to take account of all the possibilities when investing in cryptocurrency.

There are several ways to keep your cryptocurrency asset safe, secure, and free from theft.

· Due enough research on exchanges.

· Store cryptocurrency safely in a wallet. There are

different cryptocurrency wallets. Therefore, you need to take into consideration their features, security standards, the technology used, and advantages before making a choice.

· Use a hybrid strategy. It is important to separate the public and private key for deposit boxes where cryptocurrency is stored because hackers look for online wallets that are gaining popularity because they are investment options. Use an offline wallet for cryptocurrency storage, and only a little amount should be kept online.

· Using a strong password. Never repeat a password that has been used for another account. Always keep the password strong and different, and always use two-way verification before logging in. Passwords can be changed after a few months.

· Use trustworthy wallets.

· Keep the key secret. The secret key is used when sending and receiving digital currency. The key is to keep it secretive and not to be disclosed to anyone, and it should be stored somewhere safe.

· Invest in buying a cryptocurrency hardware wallet.

· Stay away from cryptocurrency gambling site.

FOLLOWING THESE MEASURES WILL HELP TO CHANGE THE narrative of cryptocurrency as not being secure and avoid loss due to theft and hackers.

Myth 4: Cryptocurrencies are Bad for the Environment

The mining process of cryptocurrency requires a lot of energy and consumes electricity to perform computational tasks of creating new coins. People tend to believe that this has a negative effect on the environment. It is observed that mining cryptocurrency generates a lot of heat, thereby warming up the surrounding space. Many argue that cryptocurrency does not affect the environment negatively, but it consumes more energy than other financial processes. Some cryptocurrency is self-dependent; as a result, it will become hard for users to generate more units over time, which will, in turn, decrease the amount of energy used to mine new coins.

There are other environmental-friendly Bitcoin alternatives, like Ethereum, Nano, Stellar, etcetera, that use modified forms of the traditional Proof of Work (PoW) mechanism.

Cryptocurrency is not the only technology that consumes power. The modern financial and banking system also uses electricity to operate on a daily basis.

MYTH 5: CRYPTOCURRENCIES ARE SCAMS

There are so many scams and frauds that revolve around cryptocurrencies, and a lot of people tend to create a belief about this that might be wrong. Cryptocurrency is not a scam. That is why it is important to have enough information about how the cryptocurrency and how the technology behind it works before investing in it.

Many people have invested in cryptocurrency but had a bad experience with it because of lack of information. It requires you to have enough information because cryptocurrency is more volatile than any other investment. This will help to maximize and minimize loss.

Many people invest in cryptocurrency without making an effort to research and get the necessary information about it. Consequently, this results in loss and unfavorable experiences. Having enough information will help to prevent numerous coin offerings, which might be one of the vices of fraudsters.

However, the risk in cryptocurrency investment is not limited to just investment. Every investment is a risk; the same way it is for every other opportunity in the investment world. All investments require thorough research, learning about the details of the opportunity, and being skeptical before investing. This will help to reduce the chances of loss and to fall into the hands of fraudsters and cybercriminals.

The essential thing required is having the right information and doing research. Unfortunately, with the big fuss about cryptocurrency, most people just invest their money blindly without doing enough research and having the right information to sustain them while investing. When the essential thing is lacking, it results in experiencing loss, and then this creates the myth about cryptocurrency being a scam.

Bitcoin has made a tremendous hit in the past few years, making people rush into investing without having the necessary information or beforehand knowledge about what they want to go into.

They become the victim of loss because they do not know when to buy and when to sell. They have no information on the effects of long-term and short-term investments. They also become a victim of online scammers because they do not know the right source to get the right information.

This has made people think that cryptocurrency is a scam, which it is not. The fact to counter this myth is to be equipped with the right information through thorough research.

MYTH 6: CRYPTOCURRENCY IS A GET-RICH-QUICK SCHEME

One of the most common myths about cryptocurrency is the belief that people have about it being a get-rich-quick scheme. This is *so* untrue. Almost everywhere on the internet is an advertisement about cryptocurrency making you a millionaire in a few days, and people become a victim of this wrong information.

It is untrue that cryptocurrency makes people rich overnight. Just like investing options like gold, mutual funds, or commodities, it takes time to yield. There is a need to weigh the pros and cons before investing. There is no investment that can make one rich in a twinkle of an eye. Depending on the type of cryptocurrency investment you want to go into, either short-term or long-term investment, it all requires taking a risk.

There are risks involved. It takes time for every investment to flourish, the same for cryptocurrency. It took Bitcoin until the last two years to substantially grow in value, it'll be the same with other coins. There is a short- and long-term investment in cryptocurrency. If you are planning on a short-term investment, there are high chances of loss because of its volatility.

Getting rich with cryptocurrency is a process that takes years. It is not a get-rich-quick scheme that everyone believes it is. With respect to that, there have been cases of scams and people getting disappointed at the way it turned out to be.

There are risks involved in investing in cryptocurrency. However, you are likely to make more money with low risk if it's a long-term investment, and high risk if it's a short-term investment.

MYTH 7: CRYPTOCURRENCY IS A BUBBLE AND CAN FADE AWAY IN THE FUTURE

Cryptocurrency has been in existence for a decade now. There have been a lot of misconceptions about cryptocurrency fading soon. Cryptocurrency will continue to exist, despite whatever challenges it might face. The advantages of cryptocurrency outweigh its critics. However, industries and companies continue to use it as their mode of payment, and individuals keep investing. The more the use of cryptocurrency, the more it gains ground in this age.

Cryptocurrencies may or may not persevere as a means of investment, but they act as an agent to change and transform the financial world and the view about money. As cryptocurrency technology grows, there is competition from private currencies that have triggered central banks in most countries to develop a digital version of their currency. Countries like the Bahamas and Nigeria have designed a digital version of their currency, while countries like China, Japan, Sweden, and some other countries are planning to launch their digital currency soon but still conducting experiments on its viability.

The idea about cryptocurrency fading away will likely not come through because every transaction, even as little as buying groceries, will soon be processed using the cryptocurrency platform. Since cryptocurrency operates on a decentralized system, where there is no need for a third party in making payments or carrying out a transaction, there is a tendency that the government might want to leverage this system but enforce contractual obligations and property rights to its use.

The advent of digital currency has helped to solve a lot of financial problems. Most organizations have adopted cryptocurrency as a means of exchange for their goods and services. In addition, some countries, like China, Sweden, the U.S., and other countries have started doing research about cryptocurrency and incorporating it into their financial system. Thus, the more individuals and organizations use cryptocurrency, the more grounded it becomes.

Over time, almost all transactions will be carried out using cryptocurrencies, from buying food and groceries at stores to paying salaries to workers, and other bigger transactions will be done with cryptocurrency.

Aside from everyday transactions, the fact that cryptocurrency is faster and easier, with no barrier to location, makes it the future. People are more likely to embrace technology because of the ease that comes with it.

However, the cost of carrying out a transaction with cryptocurrency is less than the other financial institutions, especially when the transaction is from one country to another.

MYTH 8: CRYPTOCURRENCY WILL DISPLACE THE DOLLAR

Though, it has been argued that cryptocurrency could end the use of the dollar because of its threat to the greenback's supremacy, cryptocurrency is not backed up by anything, government or authorities; it just exists based on the faith of the people who created them that it will solve a problem. Nevertheless, the dollar is backed up by the U.S. Government, and a lot of people tend to trust the dollar more than digital currencies, even when the dollar is not favorable.

There are new cryptocurrencies coming up, which are called stable coins. These stable coins aim to have stable values, which makes it easier to conduct digital payments. One of these is Diem that Facebook plans to launch, which will be backed by the U.S. dollar, giving it a stable coin. However, the value of stable coins comes from their backing by government-issued currencies.

The result of this is that dollars might become less used in making payments, but the use of the U.S. dollar as a store of value will not be tampered with.

The U.S. dollar is one of the recognized and acceptable currencies in the world. With the rise in the use of cryptocurrency, there is no likelihood that cryptocurrency will replace the dollar. The dollar is a fiat currency that has the backup of the government, and the government regulates it. Cryptocurrency is a decentralized system with no government backup, and there are no authorities to regulate its activities; the world's economy might be in displacement if there are no regulations governing the financial activities of different countries in the world.

MYTH 9: CRYPTOCURRENCY WILL REPLACE FIAT CURRENCIES

There has been some untrue information about cryptocurrency replacing fiat currencies. Despite the fact that people think cryptocurrencies are not secure and people use them for illegal purposes, there are still some who believe that it is the future and can replace paper currencies when it becomes stable.

Most industries now accept cryptocurrencies as a method of payment. Elon Musk recently announced that Tesla cars can now be bought using Bitcoins. This is a clear view that cryptocurrency is not used widely as a method of payment, but it can't replace fiat currencies.

Fiat money is a government-issued currency that is not backed up by a commodity. It gives the central bank control over the economy. Examples of common fiat currencies are the euro, pound, sterling, and dollar.

MYTH 10: CRYPTOCURRENCY TRADING IS TOO COMPLEX

Most people have little or no knowledge about cryptocurrency and trading in cryptocurrency. Nonetheless, they go about with secondhand or hearsay information, which might not be true. Some people have this misconception and do not trade in cryptocurrency because of the information they have, as being complex to operate. The truth is that cryptocurrency is as easy as buying a share on a share-trading platform. The only requirement needed is to open an account to transact. Before making transactions, research and have enough information about what you are going into. It is not as difficult as what has been perceived.

Similar to this myth is that there are rumors that cryptocurrency has gone far; there is no space for anyone to invest again. Depending on the money available for investing, there are lots of coins that you can afford to buy after considering the type of investment you are willing to go into. Cryptocurrency trading has space for everyone; it is never too late. Though Bitcoin started in 2009 and its value has risen above $50,000 recently, this doesn't mean that it can't be purchased. The important thing in cryptocurrency trading is information.

Cryptocurrency is actually easy to operate. It is just as easy as registering and opening an account.

Another important thing to note is having the information required to start-up or invest in cryptocurrency. This will help to save a lot of stress that you might encounter while trying to start.

Moreover, cryptocurrency can be invested at any time. There is no particular time where one can invest. The myth is that cryptocurrency has gone far, and there is no room for anyone to invest and buy digital assets again, but there is space for investors who want to invest in cryptocurrency. There is no limitation.

These myths and misconceptions about cryptocurrency are due to the misinformation that surrounded Bitcoin when it was first launched publicly in 2010. Society did not trust this new virtual system of transaction because of its anonymity, because of the fear that it could disappear into the thin air. Instead, it has evolved and grown massively over the years. Many still believe that cryptocurrencies are counterfeit. The truth is that cryptocurrencies cannot be counterfeited.

There are lots of cryptocurrencies in the world, and they have unique codes in which they operate. Blockchain technology is the foundation in which cryptocurrencies are being built, which makes it impossible to record the transaction a person performs and cannot be replicated. This means that they cannot be counterfeited, because the system will detect any activities that might infiltrate into the system and will crash them immediately.

Though cryptocurrencies are not government bound, this does not stop their operation, as most countries are now duplicating their use by designing their digital currencies. It is true that most people have embraced the new financial technology, but only a few have the right information about what cryptocurrency is and what it is not. Many argue that it is invisible, intangible, and it is just a product of unfamiliar technology that only a few people can relate to. The tales surrounding the history of cryptocurrency are a system where cybercriminals, fraudsters, and hackers operate. Having the right source of information about the truth and untruth of cryptocurrency will help to clear the doubt about its effectiveness and reduce the rate in which crypto traders are being scammed.

Some of these myths about cryptocurrency have been debunked and also the truth about cryptocurrency has been revealed to an extent. Lack of knowledge has caused a lot of these myths. Some beliefs are far from the truth that people tend to keep them, which has affected their trading in cryptocurrency or has affected their perception about cryptocurrency.

8

MINDFUL INVESTING IN CRYPTOCURRENCY

Cryptocurrency is a highly volatile, speculative, and risky investment that you need more than enough knowledge on before you can invest. What that means is, if you are thinking of investing, there are a few tricks you must deliberately learn to keep your money safer. Also, there are many factors to consider to determine whether you should trade or not. More so, you also need to know which wallet and marketplace you can use to invest with and maximize profit.

As I said earlier, if you are skeptical or on the fence on whether or not you should invest, you are not alone. According to a report from cryptocurrency exchange Gemini, only 14% of U.S. adult Americans own crypto accounts. It was stated that 63% of Americans are classified as "crypto curious." This is a 2021 report, which implies that over 40% of American adults are not curious about cryptocurrency, despite the hype.

The reason for lack of curiosity among some people is not far-fetched. Cryptocurrency has many drawbacks that can make someone with accurate knowledge of it lose interest.

While you can lose money in the market, you can also rely

on the platform to directly expose the demand for virtual monies. It is a good investment that provides a lucrative alternative to other traditional investments you can ever think of in the financial market.

HOW SAFE IS CRYPTOCURRENCY?

Crypto is a high-risk, high-reward kind of investment, meaning it has a lot of risks attached. It potentially is extremely profitable. So, it is an investment that offers you a huge benefit, but you may have to consider some of the risks attached to it before you venture into it.

Stocks have been around for a while, and it has gained recognition and government acceptance based on their real-life uses. But crypto is a new decentralized phenomenon that does not enjoy the same privilege with stocks and other traditional investments. However, it could become mainstream and change the world. And investors who traded early will amass wealth for themselves through the currency.

Beyond that, you must consider your risk tolerance level before you invest. If you are always worried about losing money or are so concerned about the profit you make, digital currencies may not be the best for you. In other words, you should not bother to invest in the market if you are a risk-averse investor. Not because the cryptocurrency is that bad, but it involves gaining and losing money. And you may not be able to determine when you will gain or lose money. The activities in the market will decide that.

On a similar note, digital currencies can lose more than half of their value with a timeframe and gain it back within a certain period. For instance, Bitcoin lost more than 80% of its total value in the past. The same goes for Ethereum. The market once lost 95% of its value within a year. If you will likely be at unrest or lose sleep when the market loses its value, digital currencies are not for you.

However, if you know that you can stand a fair amount of risk to maximize potential benefits, crypto may be the best fit for you. Most especially now that digital cash prices are rebounding, is it the right time to invest? Below are a few things you need to know.

HOW TO INVEST SECURELY

Despite the volatility of cryptocurrency, you can amass wealth for yourself if you know how and when to trade in the market. But first, you have to make the decision that you really want to invest. After making that decision, you should strategize on how you want to implement that decision.

The first strategy to consider is to invest digital cash that you can afford to lose. It's important you understand that, when it comes to investing, there are never any guarantees—you can't be sure whether you will gain or lose. Therefore, you should avoid investing more than you can afford from your portfolio. You probably know why. The market carries more risk than the average stock, and it's good to be on the safe side.

In addition to that information, always ensure you have at least three to six months' worth of savings set aside in your emergency fund account. This is because, as you know, the market's volatility makes the price unpredictable, and because of that, there is a chance that prices can fall beyond expectation after you have invested. If there is an urgent need to make some unexpected expense, the least decision you should probably think about is selling your currency when the price hits bottom. You can keep your money invested until prices bounce back when you have more money in your savings.

HOW TO INVEST IN CRYPTOCURRENCY

As you might know, you can't go to any financial institution or even a brokerage firm to request digital currency. This is because it's something far off to the sphere of financial institutions. Apart from the fact that the concept is misunderstood, its unregulated or decentralized nature would not allow financial institutions to accept it. For that reason, digital currencies tend to function within their capacity using their network.

Carefully read the tips below if you really want to invest.

Use Only a Small Portion of Your Portfolio for Trading

Because of the market's volatility, you will have to decide how much of your portfolio you would like to invest. Now that the price of a token is inching high, you may have to decide how much percentage of your portfolio you would like to invest.

With these advances, particularly in Bitcoin, it cannot be easy to make a decision. As you know, all investments are ruled by the combination of greed and fear. Naturally, you would want to make a profit, and at the same time, you will be afraid of losing money. It may be challenging to keep the greedy part in control, given the exploits that people are making in the market in recent times.

Notwithstanding, first, always ensure cryptocurrency takes only a small percent of your portfolio. In order words, invest only a tiny portion of the total amount you have in your portfolio. You can decide the exact amount you would like to invest. But don't allow your investment percentage to exceed 10% or 5%. This is the best and suitable percent that should be invested.

Second, understand that, though stocks are an investment,

it is not your Bitcoin or any other currencies. Much like investing in silver and gold, they don't come with an interest or dividends. The degree that the currency will be regarded as a good investment depending on the price increase insignificantly.

Third, digital currencies are not designed to be an investment. This is because there are many mediums of exchange. They have been thought of as an alternative to euros, dollar, yen, and others. Also, it's been thought that they will represent a more efficient means of commerce, especially on the web. That is because its value is determined by the market and not by the manipulation of sovereign currencies.

But contrary to the above, cryptocurrencies haven't actively and satisfactorily fulfilled the role as a medium of exchange. This is because only a minimal number of traders accept it as the medium of exchange between them and others.

Until now, both the current uses and the future possibilities of digital currencies are uncertain.

Choose from Numerous Currencies

One of the major complications of digital currencies is that we have a good number of them. We have more than a thousand of them in operation that you can choose from, with a good number of them being designed every day to engage investors.

Because there are many of them, it has caused some complications that more are coming online almost every day. This helps to make it challenging to strike a balance between those who have come and gone. And the whole idea of cryptocurrency started less than two decades ago.

Cryptocurrencies that you can choose from include Bitcoin, Ethereum, Litecoin, Zcash, Stellar Lumen, and others.

You can choose from any of them and open a wallet account on it.

Select a Platform to Buy Digital Money

One of the major drawbacks of digital currencies is that you can't find them at usual financial institutions. For the most part, you are limited to buying, holding, trading, and selling. And all these can only be done on dedicated crypto exchanges. Some of the largest of them include the following:

- Coinbase
- eToro
- Trust Wallet
- Bitpay
- Unifimoney
- Gemini
- Binance.US
- Wealthfront

Let's consider each of them accordingly.

Coinbase. This platform was designed and started its operation in 2012 with a radical approach that makes it easier for people to send and receive Bitcoins from wherever they may be. Today, the platform offers a trusted and easy-to-use system for accessing the broader crypto economy.

eToro. This is another reliable and advanced crypto investment system that offers distinct investment services with its CopyTrader technology. This design enables you to copy the most successful crypto traders' investment techniques without being charged or sanctioned. It offers CopyPortfolios that are designed like crypto Robo advisors. They are majorly used for professional portfolio management services. It also allows you to learn the elementary things about investing in Bitcoin. More so, this platform offers virtual portfolios where you can trade up to $100,000 in a paper account without risking your money.

Trust wallet. It is a multi-coin crypto wallet where you can send, receive and store cryptocurrencies and digital assets like bitcoin, Ethereum, Binance coin, and many others safely. It is the official crypto wallet for Binance. There is no restriction to what coin should be purchased on a trusted wallet, and it provides a secure system where you can buy and store multiple cryptocurrencies, as many as you want on your wallet. Binance acquired the trust wallet in 2018. Trust Wallet allows for an exchange of coins with a single account. It also allows for staking on proof-of-stake coins such as Tezos, Cosmos, or Tron on the wallet.

Bitpay. It is another type of crypto wallet used to secure a bitcoin payment service for thousands of bitcoin users. It was created alongside the bitcoin currency. It was designed to provide a faster, more secure, and less expensive system on a global scale which is allowed in over 38 counties. However, it is one of the largest bitcoin payment processors in the world. Bitpay allows businesses to accept payments in bitcoin and bitcoin cash, also sending funds directly to a bank account.

However, Bitpay uses modern technology to convert bitcoin into any fiat currency while protecting customers from volatility risk, and hence, they get direct deposits into their bank accounts.

GEMINI. THIS IS A UNIQUE PLATFORM THAT OFFERS SERVICES beyond trading cryptocurrencies. The platform provides you with a system where you can store your crypto in a digital wallet. In addition to earning up to 7.4% APY interest on any currency you select, they also offer a wide range of crypto research and tools that you can leverage for a better trading experience.

Unifimoney. With this platform, you can keep all investments and banking in one safe place. Besides that, you have access to trade and store more than thirty-three currencies through their partner crypto exchange Gemini.

In case you want all your accounts to be in a single place, or you are tired of having different applications for every trading and banking purpose, then use Unifimoney. It is a platform you can rely on when your goal is to merge all your accounts and run them through an application.

Binnance.US. This is the American version of Binance. It is cheap and easy to operate. However, the platform doesn't offer as many trading pairs between diverse cryptos as its parent company. But it is still part of the other leading exchanges. It is a platform that allows you to use your bank to deposit. This is possible through an ACH deposit limit of $5,000.

Wealthfront. If your goal is to learn and explore the market without buying or trading crypto, Wealthfront is the best fit for you. You can explore the market using two diverse means—Grayscale Ethereum Trust (ETHE) and Grayscale Bitcoin Trust (GBTC). In addition to that, you still maintain your access to a large selection of ETFs. When all these benefits are combined, you have access to more investments than ever before.

Save and Store your currencies

Typically, currencies are stored in a wallet. This wallet could either be cold or hot.

As said before, a crypto wallet is a software program that stores public and private keys that link you up to the blockchain where your currencies exist. They don't necessarily store your currency, but they give you access to the blockchain where both your private and public keys are stored. These two keys are a necessity for any transaction to be completed. We refer to them as keys because they unlock your currencies on the blockchain.

In addition to allowing you to send and receive crypto, a digital wallet also records all the transactions you have made in the blockchain. It also shows your current balance.

DIFFERENT TYPES OF DIGITAL WALLET

There are different types of crypto wallets:

Online Wallet. This type of wallet exists on a cloud, and it can only be accessed from any computer. They are easy and convenient, but your private keys are accessed and controlled by another. This is why many investors don't really like to use it. It is less secure.

Desktop Wallet. This type of wallet is installed on a personal computer. And since the information is on your system, they are secured because no third party has access to it.

Mobile Wallet. This is a type of wallet you download on your mobile phones and tablets. They have the advantages of being used anywhere the currencies are accepted.

Hardware Wallet. This is a wallet that stores your private keys on a hardware device or substance, such as a USB drive. Unlike the online wallet, their security is a guarantee because they don't store private keys online. Instead, they are stored on a device to where third parties don't have access.

SECURING YOUR CURRENCIES

Keeping your currencies safe is the best thing you can ever do as an investor. This is more needed if you want to purchase products or when it is on the hot wallet.

Basically, anytime you use crypto online, always ensure your investment is secure. You may leverage a VPN like NordVPN to ensure your transaction is not being revealed to anybody. Using a VPN helps to secure your transactions and also have your data encrypted. With that, no one will see any of your transactions online. In addition, there is an extra layer of protection that VPN gives your data and wallet, making them completely anonymous. With that, hacking becomes difficult, especially when you have a lot of currencies in your account.

In summary, investing in cryptocurrencies is the best investment plan you can ever make at this time when their values are increasing daily. When you are planning to invest, carefully consider the tips discussed above and ensure you internalize every bit of them. Also, ensure you understand different types of digital wallets and always choose the best for trading.

9

BLOCKCHAIN & CRYPTOCURRENCIES

Don't invest in what you don't know. Learn first, then invest.

— ROBERT T. KIYOSAK

Perhaps one of the dilemmas of investments is that you may have to first learn about them. This is true, especially if you are not familiar with the specific type of investment. For example, since Bitcoin broke through to the mainstream in 2013, the new asset class of cryptocurrencies has emerged. It has undergone multiple quantum leaps in less than one decade and become a force to be reckoned with, especially in places like Canada, the European Union, Canada, and the United States.

In the US alone, more than 2,300 companies now accept crypto, although there are a few governmental restrictions to regulate any money laundering possibilities. However, even the news about crypto restrictions give credence to cryptocurrency's rapid and increasing buy-in and popularity worldwide.

Things are moving fast, and you may feel left behind already. The chances are that you do not even understand the technology behind cryptocurrencies or blockchain, as it is called. So, relax, even amid the widespread fear of missing out on the next big thing again.

This chapter aims at helping you understand the relationship between blockchain and cryptocurrencies. By the time you read to the end, you will come away with sufficient knowledge of the topic, debunking myths and demystifying essential aspects of the subject. Also, this will help you create a strategy on how to enter into the crypto market as a crypto-literate person and potential user.

WHAT'S ALL THE FUSS ABOUT BLOCKCHAIN?

Experts of cryptocurrency cannot mention the history of this intelligent innovation without the genius of Satoshi Nakamoto. Like every human, Satoshi Nakamoto imagined that every person would be free from the government's control of financial resources and economic policies that restrict the average person from achieving true freedom. This is how blockchain technology began.

It brings together different people from diverse backgrounds under their passion for decentralizing governmental control. To support this point, consider this quote by Word-Proof founder Sebastian van der Lans, a big follower of Brendan Blumer, a young and famous cryptocurrency thought leader and entrepreneur.

"Everything is on the blockchain, every piece of content, every interaction. And it lives in a completely decentralized capacity, in which nobody can shut it down, nobody can take control of it,

and it is accessible by everybody. It removes the ability to censor data completely."

THAT QUOTE REPRESENTS THE GOAL OF BLOCKCHAIN TECHNOLOGY—liberty and accountability. It also serves to answer how exactly the inventors of this world-changing technology can conceal their identities and tracks, which speaks of absolute control over what they will choose to reveal to the general public. Therefore, it was only ideal for an invention like cryptocurrency to be built on it. But before you read details on that, you may want to consider how exactly blockchain works.

BLOCKCHAIN: WHAT IT IS AND HOW IT WORKS

Every person wants to know the precise meaning of a word. Take the word *computer,* for example. Understanding begins to dawn in a person's mind when you tag a computer as *an electronic device.* That is something to hold on to. "But what type of electronic device is it?" the person may ask out of curiosity.

At that point, you may have to provide what set of activities that electronic device carries out, the way it functions, and the parts it has to function in that peculiar way. These will identify the marked differences between that word you are defining and other things that look like it. In other words, your description will distinguish *a computer* from other devices that are in the category of electronic, just like itself but have different functions from it. This is the approach we will use in our definition and description of blockchain.

BLOCKCHAIN IS A PARTICULAR KIND OF ELECTRONIC DATABASE. An electronic database is a collection of information that you can organize for search to retrieve any data or information. That means that, from there, you can store, modify, or delete that information as you please or as you can in case of laws apply.

Blockchain stores information differently. It stores information and data in blockchains that are then "chained" or joined together. A blockchain assembles data in "blocks," which are sets of groups of data or information. And as new data is entered, it comes into a new block. When these data are chained together in different chronological sets, you arrive at the reality of the blockchain.

Blockchain stores different kinds of data. It has been chiefly used as a transactions ledger. For example, encrypted details on sending decentralized digital money from one party to another, such as the date and time, amount, and address of the parties.

Blockchain is decentralized. By this, we mean that there is no monopoly of control. In other words, a single person or exclusive group does not control how banks of world governments control the flow of financial resources by printing their currencies. Instead of that, blockchain has collective governance in the sense that every user has control.

Blockchain data is irreversible. Whatever data you enter as part of your transaction is irreversible on decentralized blockchains. This means that every transaction that you are involved in is recorded into permanent documents. Thus, although anonymity is somewhat guaranteed, it is still viewable to everyone on the blockchain.

Blockchain runs 24/7. Unlike banks and central authorities that may take up to a few days to settle transactions due to their protocols, blockchain works every day of the year and every moment of the day. So, while it is possible for you not to see a deposit that you initiated before the weekend unless the next weekday comes, you can finish a blockchain transaction in under ten minutes. What's more, you are certified secure after some hours, unlike the regular banking depositing system.

Moreover, the cross-border nature of blockchain makes it possible for merchants, service users, and their clients to transact business faster. This is a welcome development compared to the regular protocol where every party has to confirm the payment processing to initiate and close out the transaction successfully.

Blockchain is potentially applicable in other non-financial and non-commercial situations such as voting.

BETWEEN BLOCKCHAIN AND CRYPTOCURRENCIES

By now, you already understand that a blockchain is a technology that makes the sharing of digital funds possible in a peer-to-peer network. It hosts these transactions in a way similar to how operating systems make it possible for us to use computers and mobile phones.

Building on that identified foundation, it is easy to identify that cryptocurrencies are the means of exchange in which Satoshi Nakamoto created blockchain. Again, consider thinking of the physical dollar and imagine its virtual or digital variant, and then you understand what a cryptocurrency is. The only distinction here is that new cryptocurrencies that are outside the governance of world governments are those that are identified as "decentralized."

The precise reason we have to make this distinction is that world governments realize the need to move with the crypto tide. For that reason, they are digitizing their fiat currencies, which are also digital assets. In this way, there is no difference at all between the physical currencies and their virtual variants. In this sense, they still retain absolute control over its generation and distribution and can influence the stability of its valuation.

BLOCKCHAIN AND CRYPTOCURRENCIES: THE CONNECTION

Blockchain transcends cryptocurrencies, as it has been identified as a foundation of digitized currency. However, they seemed to both emerge around the same time. But first, why is it essential to make a distinction between these terms? It's simple: people use them synonymously, but they are not the same. The misconception that they are the same can easily rob thousands and possibly millions of people of the limitless potential through groundbreaking blockchain technology. And the effort of making a distinction between these two related terms and innovations will achieve further consciousness of newer discoveries, which will come by a disruption of existing markets some time to come.

It helps to recognize that blockchain is the database where every Bitcoin transaction is documented. Back in 2013, blockchain emerged from obscurity into the awareness of a global audience. It even earned its name because of the data grouping style that always generated a hash code. The bottom line is that it has always existed, but it was crypto that popularized its existence.

Fundamentally, disruption is at the core of cryptocurrencies and their parent invention—blockchain technology. As of now, the impacts of the advent of technology have revolutionized how we live and connect in only ten years.

As the precursor to further imminent disruptions, blockchain is bringing possibilities in the following ways:

Free Digital Money Markets. Speaking of revolutions in funds transfer between parties in a financial transaction, blockchain has made huge impacts. There is no more need for intermediaries like before, as free markets now have buyers and sellers within reach. Second, this system has significantly improved security in virtual financial transactions, as there is less vulnerability to cyber-attacks. Finally, compared to past times, there is better transparency between parties involved in these exchanges.

Expanded Market Share Among Businesses. There has never been a time than this crypto-age when enterprises could experience an incredible launch into acquiring a more significant market share. If you are wondering how that happens, the answer is two-fold. The first cryptocurrency helps increase the visibility of a business to new clients, just as it helps increase the involvement of the existing ones. Many studies have shown the relationship between at least a 40% increase in business profits and clients from whom they make those profits. The simple connection is that they pay with cryptocurrency, whether Bitcoin, Litecoin, Ethereum, or other cryptocurrencies. According to research, the more intriguing detail has been that the profits these establishments recorded were at least 100% compared to when these customers used their credit cards, the initially popular system.

Add to this the introduction of cryptocurrency into the payment options of eCommerce companies, and then you have the second reason. By now, there are tens of eCommerce websites that boldly and proudly feature the cryptocurrency payment option on their cart settings. This means that their clients can close out their purchases online by opting to pay for their goods using between ranges of cryptocurrency options in addition to other card transaction alternatives.

Smart Contracts. Smart contracts are programs that conduct crypto-code-based buyer-seller terms of the agreement within the context of the decentralized blockchain network. In this way, interactions between buyers and sellers can be tracked, keeping with the principles of transparency and permanence. In this automated system, your productivity increases while expenses like wasted time and financial resources vanish. When you consider what the synergy of blockchain and cryptocurrency will do for you, it is as simple as thinking of it as a system that helps you facilitate sharing any valuable resource or holding. That is not limited to shares and money, and you can rest assured of absolutely no conflict as well as your gains of circumventing middleman costs.

Logistics and Supply Chains. Blockchain is the open secret of Walmart, UPS, Maersk, British Airways, and FedEx. Companies like these are leveraging the blockchain in their supply chain endeavors for identity management, digital documentation, payment and invoicing, provenance or audit trailing, and logistics marketplace management. For example, Walmart uses its scanning system to take stock of products prepared for dispatch to their order locations.

As a result of blockchain, they can now track the products to the end receiver, from the point of dispatch to when they are in transit. Not just that, they can identify the source farm of a fruit product and its storage location in their back room.

Seeing blockchain encourages an open-source or a more accountable system, Walmart is using it to stay transparent to their clients. A case in point is that those who buy consumables can quickly identify the source of their favorite products. In turn, this will make for appropriate feedback to everyone in the chain, from the manufacturer or supplier to the end-user. Of course, their customers can readily pay in Bitcoin or any other cryptocurrency they prefer, but it is clear now that this is just scratching the surface on the benefits of using blockchain and crypto.

3 MYTHS ABOUT BLOCKCHAIN AND CRYPTOCURRENCIES

Myth #1. All blockchain data is public.

This myth goes along with the privacy scares that are not founded on correct fact-checking about cryptocurrency. It is false because, although transactions on the public blockchain are visible, transactions come with a decoupling feature for identities.

The fact is that blockchain transaction parties often take on identities that are represented by a mere string of characters. Blockchain is identity-concealing because, if transactional parties follow general privacy protocols, like avoiding being connected to anything that can identify their information, they will remain largely anonymous.

Moreover, blockchain users can store their information in a standard, secure cloud. This comes with a hash that other users cannot access or associate with the document. On the other hand, in a private blockchain context, the administration serves as the gatekeeper of any internal system, restricting access and managing the system.

Myth #2. Blockchain is impregnable to attacks.

Indeed, the blockchain technology system is currently a high wall that hackers cannot subdue, but that is only largely verifiable. The other side of the story is that they are not entirely invincible.

Consider the fact that, although transactions that you commit to a public or permissionless blockchain cannot be reversed, private blockchains are lacking what is called a consensus algorithm. Since it is an administrator who presides over the changes, private blockchains are a level of porosity. As for public blockchains, they are vulnerable if data is recorded off-chain.

Myth #3. Blockchain is superior to traditional databases.

We are at a time when blockchain is penetrating every system, from tech to finance to food and the supply chains.

However, companies still have not encountered blockchain's deployment on a use case within their sphere. The truth is that blockchain has its advantages and benefits, but ultimately, each company will have to determine if it will work for their industry or specialty in particular.

So far, you have seen the connection between blockchain and cryptocurrencies. In your mind, you may already be considering how to adopt cryptocurrency into your business or determining if it does not apply to you at all. The inevitable is that there will yet be an evolution of blockchain in spreading the influence of blockchain, especially in specific industries.

THOUGHTS ON CRYPTO'S UTILITY

TO USE OR NOT TO USE...

Cryptocurrency is a digital currency that can be used to buy goods and services, and it uses an online ledger with strong cryptography to secure information. It has become a widely recognized technology that seeks to solve the financial problem and reduces the control of the government on financial assets after a breakdown in the economies of developed countries, like the U.S.

The history of cryptocurrency can be traced as far back as the 90s, and it has been seen as the future because of the value it has accumulated over time. However, the cryptocurrency, which was primarily created as a means of exchange, just like every other physical currency, might not be able to displace the fiat currencies because of the inability of the government to control and regulate its operation. Many people who invest in cryptocurrency see it as a store of value because of the tendency to increase in value in the future.

Cryptocurrency is relevant in almost all parameters. For example, it is used to buy goods and services, store value for future purposes, and a lot of other important things to note in the financial space.

In the last ten years, making transactions has been quite

difficult with the cost and the formalities involved, especially when it is outside a particular location. Cryptocurrency helps to bridge the gap between the traditional and the modern method of making transactions. Furthermore, the use of cryptocurrency is not limited to any location; international transactions can be made without considering the cost.

The brain behind the safety of cryptocurrency is blockchain technology. Cryptocurrency is not the only product of blockchain technology, and it is used in every other area, like record keeping in the hospital, voting, and others.

However, cryptocurrency continues to be the trend because of the numerous advantages it has. Though there are disadvantages to the use of cryptocurrency, it is effective in expanding the financial space and economy of the world to a considerable amount.

ABOUT THE AUTHOR

Alex Caine is an international bestselling author of psychological thriller fiction and easy-to-read investment & financial books created to educate the common man on investment strategies so they can make sound decisions instead of hopping on trends and risk unnecessarily.

Alex sees himself as a layman's educator and guide (not a financial advisor) for the average person (like him) who wants to eliminate bad debt, leverage OPM (other people's money) and tax shelters, increase passive cash flow while building generational wealth, understand the new trends, investments, markets, and the future of money, and if they're fortunate, make some money along the way.

He shares what he knows or has experienced himself to those he's one step ahead of in some aspect of life, as he pursues growth from those ahead of him. Thus, creating a train of valuable insights passed from the top to the bottom.

When he's not writing fiction to entertain his readers (and himself), or sharing what he's uncovered, learned, and implemented with his investment and cash flow strategies, he can be found voraciously reading all things business, history, and wealth creation, attending masterminds and events, running around playing *"capture"* or *hide n' seek* with his two young boys, or traveling with his family.

ALSO BY ALEX CAINE

FICTION

Kiss Of Death

NONFICTION

Myths Vs. Facts Of Cryptocurrency

The Era Of NFTs